Agrometeorology

Agrometeorology

Editor

Stefanos Palmer

scitus academics

Agrometeorology
Edited by **Stefanos Palmer**

ISBN: 978-1-68117-244-6
Library of Congress Control Number: 2016934784

SCITUS Academics LLC,
www.scitusacademics.com
Box No. 4766, 616 Corporate Way,
Suite 2, Valley Cottage,
NY 10989

Printed in United States of America on Acid Free Paper

Preface

Agrometeorologyor Agricultural Meteorology involves the integration of climatological and meteorological data and techniques into agricultural problems, such as crop production, soil moisture, moisture stress, and migration of pests. Agrometeorology is the study and use of weather and climate information to enhance or expand agricultural crops and/or to increase crop production. Agrometeorology mainly involves the interaction of meteorological and hydrological factors, on one hand and agriculture, which encompasses horticulture, animal husbandry, and forestry. It is an interdisciplinary, holistic science forming a bridge between physical and biological sciences and beyond. It deals with a complex system involving soil, plant, atmosphere, agricultural management options, and others, which are interacting dynamically on various spatial and temporal scales. Specifically, the fully coupled soil-plant-atmosphere system has to be well-understood in order to develop reasonable operational applications or recommendations for stakeholders. For these reasons, a comprehensive analysis of cause-effect relationships and principles that describe the influence of the state of the atmosphere, plants, and soil on different aspects of agricultural production, as well as the nature and importance of feedback between these elements of the system is necessary. Agrometeorological methods therefore use information and data from different key sciences such as soil physics and chemistry, hydrology, meteorology, crop and animal physiology and phenology, agronomy, and others. Observed information is often combined in more or less complex models, focused on various components of system parts such as mass balances, biomass production, crop growth and yield, and crop or pest phenology in order to detect sensitivities or potential responses of the soil-biosphere-atmosphere system.This book entitled Agrometeorology emphasizes on the application of meteorological information and data to weather sensitive problems of agriculture.

Table of Contents

1

CHAPTER

CHANGING TRENDS AND ABRUPT FEATURES OF EXTREME TEMPERATURE IN MAINLAND CHINA FROM 1960 TO 2010

Shibo Fang [1,2,†], Yue Qi [3,†], Guojun Han [1,†], Qingxiang Li [4] and Guangsheng Zhou [1,*]

[1] *Institute of Eco-environment and Agrometeorology, Chinese Academy of Meteorological Sciences, No. 46 Zhongguancun South Street, Beijing 100081, China*

[2] *Collaborative Innovation Centre on Forecast Meteorological Disaster Warning and Assessment, Nanjing University of Information Science & Technology, Nanjing 210044, China*

[3] *Gansu Key Laboratory of Arid Climatic Change and Reducing Disaster/Key Laboratory of Arid Climatic Change and Disaster Reduction of CMA/Institute of Arid Meteorology, China Meteorological Administration, Lanzhou 730020, China*

[4] *National Meteorological Information Centre, China Meteorological Administration, Beijing 100081, China*

ABSTRACT

Studies based on the 10th (90th) percentiles as thresholds have been presented to assess moderate extremes in China and globally. However, there has been notably little research on the occurrences of high extremes of warm days and hot days (TX95p and TX99p) and cold nights and very cold nights (TN05p and TN01p), based on the 95th and 99th (5th and 1st) percentiles of the daily maximum (minimum) temperature data at a certain station in the period 1971–2000, which have more direct impacts on society and the ecosystem. The trends analyses of cool nights or warm days are based upon the hypothesis that expects a linear trend and no abrupt change. However, abrupt changes in the climate, especially in extreme temperatures, have been pointed to as a major threat to ecosystem services. This study demonstrates that (1) the mean frequencies of TX95p and TX99p increased by 1.80 day/10 year and 0.62 day/10 year, respectively, and that those of TN05p and TN01p decreased by 3.18 day/10 year and 1.01 day/10 year, respectively, in mainland China. Additionally, the TX95p and TX99p increased significantly by 50.42% and 58.21%, respectively, while the TN05p and TN01p of all of the stations decreased significantly by 83.76% and 76.48%, respectively. Finally, (2) the TX95p and TX99p trends underwent abrupt changes in the 1990s or 2000s, but the trends of TN05p and TN01p experienced abrupt changes in the late 1970s and early 1980s. After the abrupt change points, the trend of warm and hot days increased more rapidly than before in most regions, but the trend of cold days and very cold days decreased more slowly than before in most regions, which indicates a greater risk of heat waves in the future.

Keywords: temperature extreme; heat waves; global warming; climate change; extreme weather events; maximum temperature; minimum temperature

1. INTRODUCTION

An increase in the average global temperature since the mid-20th century has been observed, and the warming of the climate is unequivocal [1]. Global warming may have increased the frequency and intensity of extreme weather and climate events [2,3,4]. Compared to the mean temperature increase, changes in regional temperature extremes have more direct impacts on society and the ecosystem [5]. Extremely hot summers can drastically reduce agricultural production [6,7,8], increase energy consumption [9], and lead to hazardous health conditions [10,11,12]. Thus, understanding and predicting the spatial and temporal variability and trends of extreme weather events is crucial for the protection of socio-economic well-being and for understanding extreme weather events for mitigation of their regional impact.

To analyse the variations in extreme climate, the Expert Team on Climate Change Detection and Indices (ETCCDI) defined 27 extreme temperature and precipitation indices [13]. Two main types of extreme indices were developed: the number of days in a year exceeding specific thresholds that have fixed values (absolute thresholds) and thresholds that are relative (percentile thresholds) to a base period climate [14]. Indices of the number of days above or below percentile thresholds are more suitable for spatial comparisons of extremes than those based on absolute thresholds [13]. Thus, extreme temperature indices that are based on minimum temperatures below the long-term 10th percentile and/or maximum temperatures above the long-term 90th percentile have been widely used in published studies in North America, South America, Europe, Asia, Australia, and on the global scale [2,15,16,17,18,19,20,21,22]. The more extreme an event, the more likely it is to cause societal or environmental damage. However, most of the studies based on the 10th (90th) percentiles as thresholds are set to assess moderate extremes that, on average, occur 36.5 times every year (10 per cent of 365 days) rather than high-impact weather events that occur once or twice per year. Compared to the moderate extremes, the high extreme temperatures that are based on 5th and 1st (95th and 99th) percentiles, which occur on average 18.25 times and 3.65 times, respectively, every year (5 per cent and 1 per cent of 365 days), have higher potential risks on people's health and lifestyles, the economy, society, and the environment. However, there has been very little research on the occurrences of high extreme warm days and cold nights according to the 5th or 1st (95th or 99th) percentiles.

More than 70% of the Earth's land area has experienced a significant reduction in the number of cool nights but an insignificant increase in warm days [2]. In China however, regional analyses have reported that a significant reduction in cool nights and a significant increase in warm days have occurred [23,24,25,26]. Along with these regional analyses in China, the changes are much less spatially coherent, even though significant trends are found in more than half of the stations in the entire China mainland. More updated studies are needed to explain the spatial heterogeneity of the temporal trends of temperature extremes in the different climate regions in China.

Analyses of the increasing or decreasing trends in cool nights or warm days are based on the hypothesis that predicts linear, rather than abrupt, changes. Abrupt changes in climate factors, especially abrupt changes in extreme temperatures, have been pointed out as a major threat to ecosystems [27,28], because extreme temperature is an important physical restriction for some plants [29,30]. Several researchers investigated abrupt changes in the trends in temperature extremes in the northern hemisphere, North America, the northwest of China and mainland China [31,32,33,34,35]; for example, an abrupt change in both temperature and precipitation extremes in Northwest China occurred in around 1986 [34], but studies of the abrupt changes in temperature extremes in mainland China used the moderate extremes temperature indices based on the 10th

(90th) percentiles. The abrupt changes of high extreme warm days and cold nights according to the 5th or 1st (95th or 99th) percentiles are studied less frequently.

In this paper, two questions are addressed with respect to temperature extremes: how the extreme temperature trends are spatially distributed in different regions in mainland China and when the extreme temperature trends' abrupt changes occurred during 1960-2010. In this study, the 5th (95th) and 1st (99th) percentiles were individually considered to determine the thresholds of four indices (TX95p, TX99p, TN05p, TN01p) based on the 95th and 99th percentiles of daily Tmax and the 5th and 1st percentiles of daily Tmin, respectively. The spatial heterogeneity and abrupt features of the temporal trends of the four indices were considered across nine climate regions in mainland China from 1960 to 2010.

2. DATA AND METHODS

2.1. Data Source and Quality Control

The National Meteorological Information Centre (NMIC) of the China Meteorological Administration (CMA) developed the first national homogenized temperature data set [36] and its more recent updated version [37]. To avoid inconsistency in percentile-based indices of temperature extremes, the newly homogenized data sets of daily maximum and minimum temperatures over the period from 1951 to 2010 at 825 stations were chosen for this study. Due to the fact that much daily data was missing at some stations, especially in the years prior to 1960, we chose a time series from 1960 to 2010. Further, according to the criteria that the series length should be no less than 51 years and that the missing data should be no more than 2% of the data points in every year at a station, the data of 591 stations over the period from 1960 to 2010 were ultimately selected for analysis. In the data sets of these 591 stations, a total of 34,776 missing daily data, accounting for 0.316% of the total data, were found at 95 of the 591 stations in the period from 1960 to 2010. The largest numbers of missing data were less than 2% in every year for all stations, so the missing values were ignored in the following analyses. The 591 stations that had good-quality data were chosen for the analysis (Figure 1).

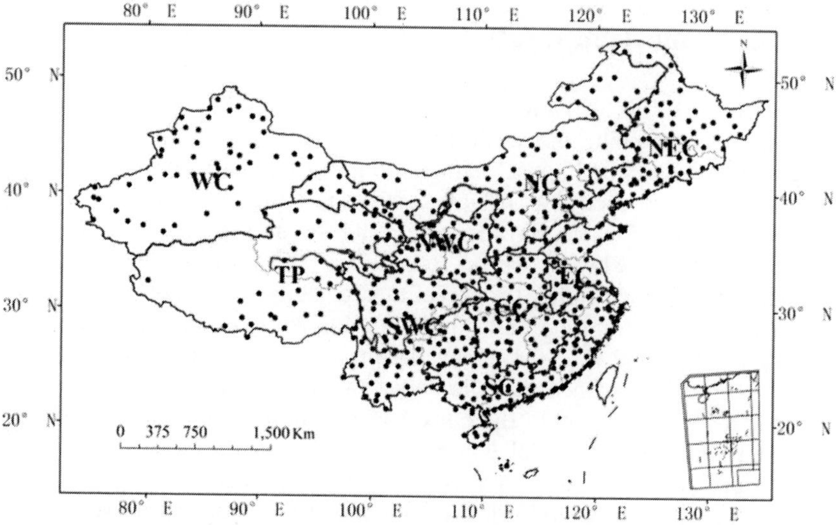

Figure 1. Distribution of the weather stations and climate zones. The upper curved purple line is the Yellow River and lower curved purple line is the Yangtze River. The blue line is used to separate the different climate zones. See text for climate zone definitions.

2.2. Climate Zones in China

The spatial heterogeneity of the temporal trends requires the definition of climate zones in China. Based on the monthly temperature and precipitation data, and considering the coincidence with the administrative divisions of provinces, mainland China was divided into nine climate zones (Figure 1) [38]: Northeast China (NEC: Liaoning, Jilin, Heilongjiang), North China (NC: Beijing, Tianjin, Hebei, Shaanxi, Inner Mongolia), Northwest China (NWC: Shanxi, Gansu, Ningxia), East China (EC: Shanghai, Jiangsu, Zhejiang, Anhui, Shandong), West China (WC: Xinjiang), Southwest China (SWC: Chongqing, Sichuan, Guizhou, Yunnan), South China (SC: Guangdong, Guangxi, Hainan, Fujian), Central China (CC: Jiangxi, Henan, Hunan, Hubei) and Tibetan Plateau (TP: Xizang, Qinghai).

2.3. Extreme Temperature Indices

We used the 95th (5th) and 99th (1st) percentiles individually to obtain the thresholds of the four indices (Table 1). The indices are different from those recommended by the Expert Team on Climate Change Detection and Indices (ETCCDI), which used the 10th percentile to define the occurrences of cold nights (days) and warm days (nights). The thresholds for the temperature extremes at each station are set to the 95th and 99th percentiles of daily Tmax and to the 5th and 1st percentiles of daily Tmin. We calculated the percentile values and defined the thresholds with respect to the period of 1971–2000 [18,39,40]. In the calculation of these indices, the 95th (5th) and 99th (1st) percentiles of the daily maximum (minimum) temperature data at a certain station in the period 1971–2000 were taken as the upper (lower) threshold. In calculation of percentile indices, a bootstrap procedure is used to avoid a bias between the percentiles within the base period and those outside [39]. If a daily maximum temperature was greater (less) than the upper (lower) threshold, then it was considered a warm (cool) day (night) event. The amount of missing data was less than 2% at all the stations, so the missing values were ignored when the percentile thresholds were estimated. According to the percentile thresholds, the temperature indices are classified as cold-related indices TN05p (cold nights) & TN01p (very cold nights) and warm-related indices TX95p (warm days) & TX99p (hot days).

Table 1. Definition of four temperature indices in China.

Name	Index	Definition
Warm days	TX95p	Count of days where maximum temperature TX > 95th percentile
Hot days	TX99p	Count of days where maximum temperature TX > 99th percentile
Cold nights	TN05p	Count of days where minimum temperature TN < 5th percentile
Very Cold nights	TN01p	Count of days where minimum temperature TN < 1st percentile

2.4. Trend Analysis, Significance Test, and Abrupt Change Detection

The annual time-series of the indices were calculated for each station. The trends in the annual indices were calculated using linear trends estimated by linear least squares for each station from all available years from 1960 to 2010. The statistical significance of the trends was evaluated at the 5% level of significance against the null hypothesis.

The possible abrupt changes in the trends of the indices were examined using the Mann-Kendall method and the 5-year moving T-test (MTT) method [17,41,42,43]. For the emergence of multiple point mutations after using the Mann–Kendall method, the 5-year moving T-test method was used to verify the authenticity of the mutation point to enhance the credibility of the results [17,41]. The Mann–Kendall statistical test has been frequently used to quantify the significance of trends in meteorological time series [44,45,46,47,48]. The Mann–Kendall is calculated as [42,43]:

$$S = \sum_{i=1}^{n-1} \sum_{j=i+1}^{n} \mathrm{sgn}\,(x_j - x_i) \tag{1}$$

where n is the number of data points, xi and xj are the data values in the time series i and j (j > i), respectively, and sgn(x_j – x_i) is the sign function as:

$$\mathrm{sgn}(x_j - x_i) = \begin{cases} +1 & if \quad x_i - x_j > 0 \\ 0 & if \quad x_i - x_j = 0 \\ 1 & if \quad x_i - x_j = 0 \end{cases} \tag{2}$$

The variance is computed as:

$$\mathrm{Var}(S) = \frac{n(n-1)(2n+5) - \sum_{i=1}^{p} t_i(t_i-1)(2t_i+5)}{18} \tag{3}$$

where n is the number of data points, P is the number of tied groups, the summary sign (Σ) indicates the summation over all tied groups, and t_i is the number of data values in the Pth group. If there are not tied groups, this summary process can be ignored [49]. A tied group is a set of sample data having the same value. In cases where the sample size *n* > 30, the standard normal test statistic ZS is computed using Equation (4):

$$Zs = \begin{cases} \frac{S-1}{\sqrt{Var(S)}}, & if \quad S > 0 \\ 0 & if \quad S = 0 \\ \frac{S+1}{\sqrt{Var(S)}}, & if \quad S < 0 \end{cases} \tag{4}$$

Positive values of ZS indicate increasing trends while negative ZS values show decreasing trends. Testing trends is done at the specific α significance level. When $|ZS| > |Z1\text{-}\alpha/2|$, the null hypothesis is rejected and a significant trend exists in the time series. $Z1\text{-}\alpha/2$ is obtained from the standard normal distribution table. In this study, $\alpha = 0.05$ was used. At the 5% significance level, the null hypothesis of no trend is rejected if $|ZS| > 1.96$.

3. RESULTS

3.1. Trends of Temperature Extremes

This section gives an overview of the time series for the four temperature extreme indices averaged over all the stations in mainland China (Figure 2). Figure 2 summarizes the anomalies in the annual temperature extreme indices, averaged over all the stations in mainland China. The warm days (TX95p) and hot days (TX99p) showed increasing trends over the last 51 years, with linear trends of +1.80 day/10 year and +0.62 day/10 year, respectively (Figure 2a), which is much less than the trend of TX90p, +5.22 day/10 year [18]. The cold nights (TN05p) and very cold nights (TN01p) showed linear trends of −3.18 day/10 year and −1.01 day/10 year, respectively (Figure 2b), which is also much less than the trend of TN10p, -8.23 day/10 year [18]. Both the warm days and hot days showed rapidly increasing trends after the mid-1980s, although slight decreasing tendencies appeared before this time. Additionally, cold nights and very cold nights generally followed downward trends (Figure 2 and Figure 3).

3.2. Spatial Variations of Trends in Extreme Temperature

This section shows the spatial variations of the trends in the four indices in mainland China (Figure 4) and the percentage of the stations that exhibited trends that passed the significance test ($p < 0.05$) among the 9 climate zones (Table 2).

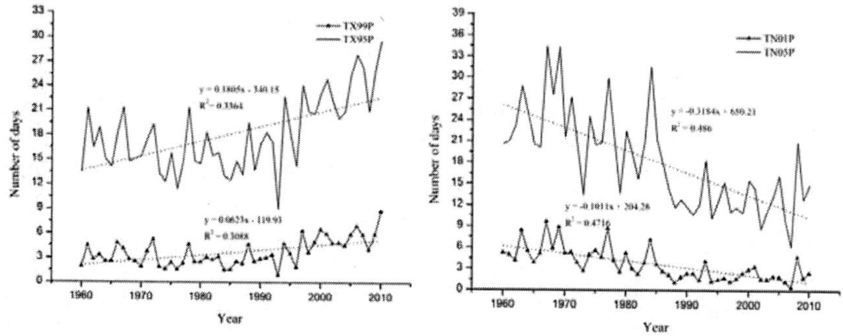

(a) (b)

Figure 2. Time series of annual occurrences of warm days (TX95p), hot days (TX99p), cold nights (TN05p), and very cold nights (TN01p) in mainland China during 1960–2010. (**a**) Warm days (TX95p) are the upper lines and hot days (TX99p) are the lower lines; (**b**) cold nights (TN05p) are the upper lines and very cold nights (TN01p) are the lower lines. The dashed lines are linear trends that are statistically significant at *p*-value < 0.05.

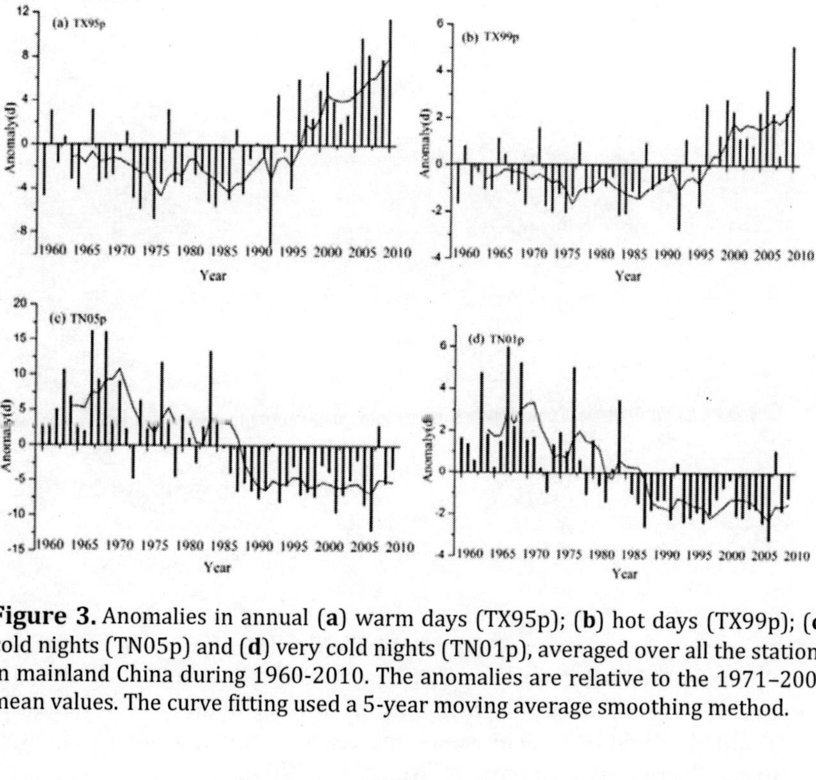

Figure 3. Anomalies in annual (**a**) warm days (TX95p); (**b**) hot days (TX99p); (**c**) cold nights (TN05p) and (**d**) very cold nights (TN01p), averaged over all the stations in mainland China during 1960-2010. The anomalies are relative to the 1971–2000 mean values. The curve fitting used a 5-year moving average smoothing method.

Table 2. Percentages of stations that passed the significance test ($p < 0.05$) among the nine climate zones in mainland China. TX95p: warm days, TX99p: hot days, TN05p: cold nights, TN01p: very cold nights, NEC: Northeast China, NC: North China, NWC: Northwest China, EC: East China, WC: West China, SWC: Southwest China, SC: South China, CC: Central China, TP: Tibetan Plateau.

	TX95p		TX99p		TN05p		TN01p	
	Increase (%)	Decrease (%)	Increase (%)	Decrease (%)	Increase (%)	Decrease (%)	Increase (%)	Decrease (%)
Mainland China	50.42	3.38	58.21	4.74	0.85	83.76	1.18	76.48
NEC	37.50	0	43.06	0	1.39	86.11	1.39	37.5
NC	68.67	1.20	72.29	1.20	1.20	87.95	1.20	91.57
SWC	48.89	3.33	51.11	4.44	1.11	80.00	0	100
TP	40.43	0	44.68	0	2.13	39.36	0	18.09
WC	51.02	6.12	48.98	4.08	0	81.63	8.16	91.84
SC	65.33	4.00	58.67	2.67	0	82.67	0	100
NWC	66.67	1.96	74.51	0	0	80.39	0	80.39
CC	8.70	11.59	18.84	36.23	0	79.71	0	78.26
EC	32.73	1.82	32.73	7.27	0	96.36	1.82	49.09

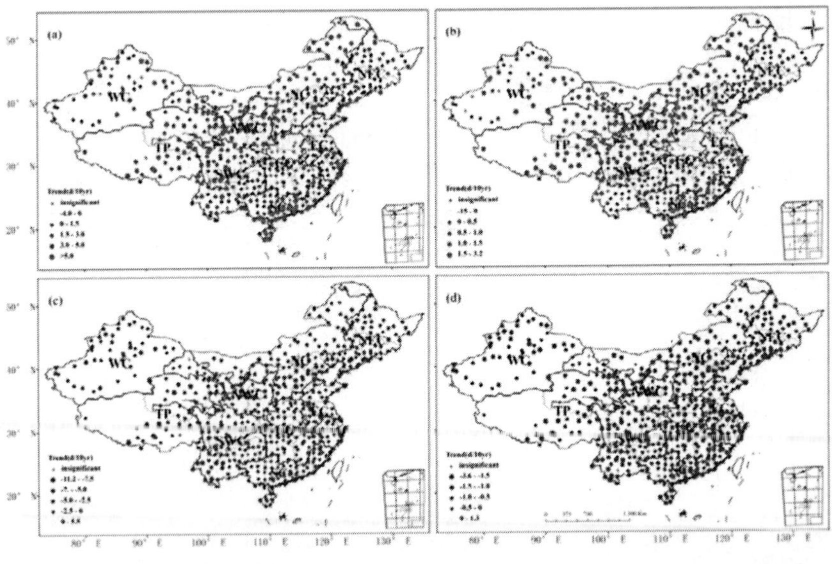

Figure 4. Spatial distributions of trends of extreme temperature indices in China during 1960–2010: (**a**) Trends of TX95p; (**b**) Trends of TX99p; (**c**) Trends of TN05p; (**d**) Trends of TN01p. The black symbol "+" shows the stations with trends that were insignificant at the confidence level of 0.05. The unfilled symbol "∘" represents increased frequency in TX95p and TX99p and decreased frequency in TN05p and TN01p.

The trends of TX99p and TX95p have similar spatial patterns, although the trend values may be different, showing an increasing trend in all regions of the mainland except for Central China (CC) and its surrounding areas, which show decreasing or insignificant trends (Figure 4a,b). This is similar to the TX90p spatial trend distribution, which showed an increasing trend in nearly all the regions except for a few stations in the central region and the eastern part of Southwest China [18]. Except for the junction area of North and Northwest China (NC and NWC), the northwest of Southwest China (SWC), and a few sites in the northeast of East China (EC), most sites exhibited decreasing trends in TN05p and TN01p (Figure 4c,d) and nearly all the stations exhibited significant decreasing trends in TN10p, with a few exceptions in the eastern part of Southwest China [18]. Both the largest increases in the TX95p and TX99p frequency and the largest decreases in TN05p and TN01p are located in the same regions of NC, NWC, WC, TP, SWC, SC north of CC, and WC, especially in the upper reaches of the Yellow and Yangtze Rivers and the estuaries of the Yangtze River.

TX95p and TX99p increased significantly in more than half of the stations, and TN05p and TN01p decreased significantly in more than three-quarters of the stations (Table 2). In mainland China, more than half of the stations exhibited significantly increasing trends in TX95p and TX99p (50.42% and 58.21%, respectively), and 83.76% and 76.48% of the stations showed significant decreasing tendencies in TN05p and TN01p, respectively. The percentages of sites that exhibited a significantly decreasing trend in TX95p and TX99p were 3.38% and 4.74%, respectively, while those showing significantly increasing trends in TN05p and TN01p were 0.85% and 1.18%, respectively.

Except for CC, where only 8.7% of the stations for TX95p and 18.84% of the stations for TX99p significantly increased and 11.59% of the stations for TX95p and 36.23% of the stations for TX99p significantly decreased, 35% of stations had significantly increasing tendencies and less than 10% of the stations had significantly decreasing trends among the other eight climate zones (Table 2). More than 80% of the stations among the other eight climate zones showed significant decreases in TN05p except for TP, where only 39.36% of the stations show significant decreases in TN05p and more than 80% of the stations showed significant decreases in TN01p, excluding TP, NEC, and EC.

3.3. Characteristics of Abrupt Changes in TX95p and TX99p Trends

The abrupt changes in the TX95p and TX99p trends mainly occurred in the 1990s and 2000s (Table 3). The abrupt changes in the trends of TX95p occurred earliest (in the 1980s) in NEC and CC and latest (in the 2000s) in SWC, SC, and EC. Before abrupt changes in TX95p, the trends were negative except for NEC, SWC, TP, and SC, but after the abrupt changes, all the trends were positive. The TX99p abrupt changes occurred earliest in SC and CC in the mid-1980s and latest (in the 2000s) in NEC, SWC, and EC (Table 3). Before the abrupt change points in TX99p, the trends were negative in NC, SC, CC, and EC, but after the abrupt change points, the trends were negative only in NC, WC, and EC.

The trends in TX95p among the climate zones after the abrupt changes were larger than before the abrupt changes, but the trend values of TX99p in more than half of the climate zones (NEC, NC, WC, NWC, and EC) were larger before the abrupt changes (Table 3).

Table 3. Years of abrupt changes and trends before and after the abrupt changes in TX95p and TX99p among the climate zones.

	Year of Abrupt Change		Trends (day/10 year) Before and After Abrupt Changes			
	TX95p	TX99p	TX95p		TX99p	
			Before	After	Before	After
NEC	1981	2001	0.7	2.9	0.9	0.4
NC	1995	1996	−0.9	1.8	−0.4	−1.0
SWC	2003	2001	0.45	12.1	0.1	5.3
TP	1992	1993	1.3	6.5	0.2	2.0
WC	1994	1996	−0.05	0.74	0.04	−0.28
SC	2000	1985	1.08	10.2	−0.5	1.7
NWC	1995	1997	−0.68	1.2	0.2	0.1
CC	1987	1987	−4.7	4.1	−1.7	1.1
EC	2000	2001	−0.07	0.27	−0.2	−1.1

3.4. Characteristics of Abrupt Changes in TN05p and TN01p Trends

Except for TP, where the abrupt change in TN05p occurred in 2001, the other climate zones' abrupt changes in TN05p and TN01p occurred in the 1980s and the late 1970s (Table 4). Before the abrupt change points in TN05p, the regional trends were negative except for WC and CC. Before the abrupt change points in TN01p, the regional trends were negative with the exception of WC. For both TN05p and TN01p, after the abrupt change points, the trends in all the regions were negative.

The decreasing trends of TN01p for all climate zones after abrupt changes were lower than before the abrupt changes, except for WC; decreasing trends in TN05p were observed in more than half of the climate zones (NC, TP, SC, NWC, and EC) (Table 4).

Table 4. Years of abrupt changes and trends before and after abrupt changes in TN05p and TN01p.

	Year of Abrupt Change		Trends (day/10 year) Before and After Abrupt Changes			
	TN05p	TN01p	TN05p		TN01p	
			Before	After	Before	After
NEC	1980	1976	−0.9	−1.7	−2.5	−1.2
NC	1980	1981	−2.6	−2.1	−1.0	−0.1
SWC	1985	1986	−0.7	−2.1	−0.7	−0.4
TP	2001	1986	−3.4	−2.7	−1.7	−1.0
WC	1980	1979	0.003	−1.2	0.67	−0.2
SC	1979	1985	−2.6	−2.3	−0.5	−0.2
NWC	1980	1981	−1.8	−1.1	−0.4	0.2
CC	1984	1980	0.23	−0.07	−0.5	−0.3
EC	1985	1983	−0.59	−0.05	−1.0	−0.6

4. DISCUSSION

The present study shows that, in mainland China, the frequencies of TX95p and TX99p increased, on average, by 1.80 day/10 year and 0.62 day/10 year, respectively, and the frequencies of TN05p and TN01p decreased, on average, by 3.18 day/10 year and 1.01 day/10 year, respectively. Zhou and Ren [18] showed that a larger increasing trend of 5.22 day/10 year occurred for warm days (TX90p) and a higher decreasing trend of −8.23 day/10 year occurred for cool nights (TN10p) during 1961–2008. The increasing rates of warm days in the 90th, 95th, and 99th percentiles are much lower than the decreasing rates of cold nights in the 10th, 5th, and 1st percentiles, respectively. This seems to be associated with the asymmetric warming characteristic in which the rate of increase in the daily minimum temperature is significantly higher than that of the daily maximum temperature [50,51,52]. Increases in the number of warm days and decreases of cold nights were also detected in different seasons or subareas of China [23,53,54,55]. The increase (decrease) in the number of warm days (cold nights) in China, whether at the 90th (10th), 95th (5th), or 99th (1st) percentiles, are consistent with all other global or regional studies that show that the occurrence of warm days increased but that cold nights decreased [2,15,16,17]. The abrupt changes in the trends of TX95p and TX99p mainly occurred in the 1990s and 2000s. The abrupt changes in TN05p and TN01p occurred in the 1980s and the late 1970s. Zhou and Ren [18] showed negative anomalies in warm days (TX90p) for most years before 1987 and cool days (TN10p) primarily showed positive anomalies before the late 1980s. The changes of TX95p and TX99p may be related to the decadal change of the East Asian summer monsoon rainfall patterns in the 1990s and 2000s [56,57,58] and the changes in the TN05p and TN01p may be related to the decadal weakening of the East Asian winter monsoon in the late 1980s [59,60].

This study showed that the frequencies of warm and hot days exhibited increasing trends, while cold and very cold nights exhibited decreasing tendencies almost everywhere except for Central China and its surrounding areas where the warm and hot days tended to decrease. In Central China, Tmax or warm days in the summer showed a cooling trend, as was also found by Qi and Wang [61] and [23]. The abrupt changes in TX95p and TX99p trends were generally in the 1990s or 2000s, but the trends of TN05p and TN01p change abruptly in the early and mid-1980s. Similar dates of abrupt changes in temperature trends were detected by Qi and Wang [61], and the trends in East Asia were found to reverse after 1997 [62]. The research shows that after the abrupt change, the increasing rates of warm days and hot days are much greater than those before in most areas of China. The number of extreme warm days (with daily maximum temperature >35 °C) increased significantly in most of China during 1960–2007 [63]. An ensemble multi-model of eight state-of-the-art GCMs(General Circulation Models) from the CMIP5 project projected more extreme warm events and fewer cold events over China in the future under the RCP4.5 scenario (Representative Concentration Pathway 4.5 which named after a possible range of radiative forcingvalues in the year 2100 relative to pre-industrial values +4.5 W/m^2) [64]. These indicate a greater potential risk of heat waves in the future, which can not only affect human health and disease but can also change the probability of agrometeorological disasters. Crop growth durations were shortened and the sowing date and phenology were shifted because of climate warming [65,66,67]. There is limited knowledge of crops' responses to climate warming, and this topic deserves more attention in the future.

5. CONCLUSIONS

(1) Warm days and hot days exhibited increasing trends over the last 51 years with a rapid increase after the mid-1980s. The cold and very cold nights exhibited decreasing trends in the last 51years with a rapid decrease from the 1960s to the 1990s.

(2) The warm days (cold nights) and hot days (very cold nights) showed upward (downward) tendencies in most areas of China, but Central China and its surrounding areas showed decreasing tendencies in warm days and hot days.

(3) The dates of observed abrupt changes in the trends of warm days and hot days were generally in the 1990s or 2000s, but the trends of cold nights and hot days changed in the early and mid-1980s. The increasing trends of warm days and hot days were greater after abrupt changes in most regions, which indicate a greater potential risk of heat waves in the future.

ACKNOWLEDGMENTS

This work was supported by the National Natural Science Foundation of China (NSFC) (No. 41375117), the International Science and Technology Partnerships Program of China and Canada (2009DFA91900) and the National Basic Research Program of China funded by MOST China (2010CB951300).

AUTHOR CONTRIBUTIONS

Shibo Fang and Guangsheng Zhou designed the research. Guojun Han and Yue Qi performed the research with significant contributions from Qingxiang Li, and Shibo Fang wrote the manuscript.

REFERENCES

1. Intergovernmental Panel on Climate Change. Summary for policymakers. In *Climate Change 2013: The Physical Science Basis. Contribution of Working Group to the Fifth Assessment Report of the Intergovernmental Panel on Climate Change*; Stocker, T.F., Qin, D., Plattner, G.-K., Tignor, M., Allen, S.K., Boschung, J., Nauels, A., Xia, Y., Bex, V., Midgley, P.M., Eds.; Cambridge University Press: Cambridge, UK; New York, NY, USA, 2013.
2. Alexander, L.V.; Zhang, X.; Peterson, T.C.; Caesar, J.; Gleason, B.; Klein Tank, A.M.G.; Haylock, M.; Collins, D.; Trewin, B.; Rahimzadeh, F.; *et al.* Global observed changes in daily climate extremes of temperature and precipitation.*J. Geophys Res. Atmos.* **2006**, *111*, D05109.
3. Intergovernmental Panel on Climate Change. *Managing the Risks of Extreme Events and Disasters to Advance Climate Change Adaptation*; Cambridge University Press: Cambridge, UK, 2012.
4. Habeeb, D.; Vargo, J.; Stone, B., Jr. Rising heat wave trends in large US cities. *Nat. Hazard.* **2015**, *76*, 1651–1665.
5. McGregor, G.R.; Ferro, C.A.T.; Stephenson, D.B. *Projected Changes in Extreme Weather and Climate Events in Europe Extreme Weather Events and Public Health Responses*; Springer: New York, NY, USA, 2005; pp. 13–23.
6. Asseng, S.; Foster, I.A.N.; Turner, N.C. The impact of temperature variability on wheat yields. *Glob. Chang. Biol.* **2011**,*17*, 997–1012.
7. Farooq, M.; Bramley, H.; Palta, J.A.; Siddique, K.H.M. Heat stress in wheat during reproductive and grain-filling phases. *Crit. Rev. Plant Sci.* **2011**, *30*, 491–507.
8. Lalić, B.; Eitzinger, J.; Thaler, S.; Vučetić, V.; Nejedlik, P.; Eckersten, H.; Jaćimović, G.; Nikolić-Djorić, E. Can agrometeorological indices of adverse weather conditions help to improve yield prediction by crop models?*Atmosphere* **2014**, *5*, 1020–1041.
9. Hadley, S.W.; Erickson, D.J.; Hernandez, J.L.; Broniak, C.T.; Blasing, T.J. Responses of energy use to climate change: A climate modeling study. *Geophys. Res. Lett.* **2006**, *33*, L17703.
10. Dematte, J.E.; O'Mara, K.; Buescher, J.; Whitney, C.G.; Forsythe, S.; McNamee, T.; Adiga, R.B.; Ndukwu, I.M. Near-fatal heat stroke

during the 1995 heat wave in Chicago. *Ann. Intern. Med.* **1998**, *129*, 173–181.

11. Pantavou, K.; Theoharatos, G.; Mavrakis, A.; Santamouris, M. Evaluating thermal comfort conditions and health responses during an extremely hot summer in Athens. *Build. Environ.* **2011**, *46*, 339–344.
12. Nitschke, M.; Hansen, A.; Bi, P.; Pisaniello, D.; Newbury, J.; Kitson, A.; Tucker, G.; Avery, J.; Dal Grande, E. Risk factors, health effects and behaviour in older people during extreme heat: A survey in south Australia. *Int. J. Environ. Res. Public Health* **2013**, *10*, 6721–6733.
13. Klein Tank, A.M.G.; Zwiers, F.W.; Zhang, X. *Guidelines on Analysis of Extremes in a Changing Climate in Support of Informed Decisions for Adaptation*; World Meteorological Organisation (WMO): Geneva, Switzerland, 2009.
14. Chen, Q.; Li, Z.; Fan, G.; Zhu, K.; Zhang, W.; Zhu, H. Indications of stratospheric anomalies in the freezing rain and snow disaster in south China, 2008. *Sci. China Earth Sci.* **2011**, *54*, 1248–1256.
15. Bonsal, B.R.; Zhang, X.; Vincent, L.A.; Hogg, W.D. Characteristics of daily and extreme temperatures over Canada. *J. Clim.* **2001**, *14*, 1959–1976.
16. DeGaetano, A.T.; Allen, R.J. Trends in twentieth-century temperature extremes across the United States. *J. Clim.* **2002**,*15*, 3188–3205.
17. Klein Tank, A.M.G.; Können, G.P. Trends in indices of daily temperature and precipitation extremes in Europe, 1946–99. *J. Clim.* **2003**, *16*, 3665–3680.
18. Zhou, Y.; Ren, G. Change in extreme temperature event frequency over mainland China, 1961–2008. *Clim. Res.* **2011**,*50*, 125–139.
19. Kothawale, D.R.; Revadekar, J.V.; Kumar, K.R. Recent trends in pre-monsoon daily temperature extremes over India. *J. Earth Syst. Sci.* **2010**, *119*, 51–65.
20. Aguilar, E.; Peterson, T.C.; Obando, P.R.; Frutos, R.; Retana, J.A.; Solera, M.; Soley, J.; García, I.G.; Araujo, R.M.; Santos, A.R.; *et al.* Changes in precipitation and temperature extremes in central America and northern south America, 1961–2003. *J. Geophys. Res.* **2005**, *110*, D23107.

21. Rusticucci, M. Observed and simulated variability of extreme temperature events over south America. *Atmos. Res.***2012**, *106*, 1–17.
22. Nemec, J.; Gruber, C.; Chimani, B.; Auer, I. Trends in extreme temperature indices in Austria based on a new homogenised dataset. *Int. J. Climatol.* **2013**, *33*, 1538–1550.
23. You, Q.; Ren, G.; Fraedrich, K.; Kang, S.; Ren, Y.; Wang, P. Winter temperature extremes in china and their possible causes. *Int. J. Climatol.* **2013**, *33*, 1444–1455.
24. Liang, K.; Bai, P.; Li, J.; Liu, C. Variability of temperature extremes in the yellow river basin during 1961–2011. *Quat. Int.* **2014**, *336*, 52–64.
25. Wang, B.; Zhang, M.; Wei, J.; Wang, S.; Li, S.; Ma, Q.; Li, X.; Pan, S. Changes in extreme events of temperature and precipitation over Xinjiang, northwest China, during 1960–2009. *Quat. Int.* **2013**, *298*, 141–151.
26. Yu, Z.; Li, X. Recent trends in daily temperature extremes over northeastern China (1960–2011). *Quat. Int.* **2015**, *380–381*, 35–48.
27. Wood, E.M.; Pidgeon, A.M. Extreme variations in spring temperature affect ecosystem regulating services provided by birds during migration. *Ecosphere* **2015**, *6*, art216.
28. Zhang, Y.; Voigt, M.; Liu, H. Contrasting responses of terrestrial ecosystem production to hot temperature extreme regimes between grassland and forest. *Biogeosciences* **2015**, *11*, 5997–6017.
29. Jeffree, E.P. Temperature-ranges and distributions of some vascular plants. *Nature* **1955**, *484*, 703–704.
30. Criddle, R.S.; Hopkin, M.S.; McArthur, E.D.; Hansen, L.D. Plant distribution and the temperature coefficient of metabolism. *Plant Cell Environ.* **1994**, *17*, 233–243.
31. Matyasovszky, I.; Ljungqvist, F. Abrupt temperature changes during the last 1500 years. *Theor. Appl. Climatol.* **2013**,*112*, 215–225.
32. Temnerud, J.; Weyhenmeyer, G.A. Abrupt changes in air temperature and precipitation: Do they matter for water chemistry. *Glob. Biogeochem. Cycles* **2008**, *2008*, 135–157.
33. Yonetani, T.; Mccabe, G.J. Abrupt changes in regional temperature in the conterminous United States, 1895–1989.*Clim. Res.* **1994**, *4*, 13–23.

34. Chen, Y.; Deng, H.; Li, B.; Li, Z.; Xu, C. Abrupt change of temperature and precipitation extremes in the arid region of northwest China. *Quat. Int.* **2014**, *336*, 35–43.
35. Liu, L.; Xu, Z.-X.; Huang, J.-X. Spatio-temporal variation and abrupt changes for major climate variables in the Taihu Basin, China. *Stoch. Environ. Res. Risk Assess.* **2012**, *26*, 777–791.
36. Li, Q.; Zhang, H.; Liu, X.; Chen, J.; Li, W.; Jones, P. A mainland china homogenized historical temperature dataset of 1951–2004. *Bull. Am. Meteorol. Soc.* **2009**, *90*, 1062–1065.
37. Xu, W.; Li, Q.; Wang, X.L.; Yang, S.; Cao, L.; Feng, Y. Homogenization of Chinese daily surface air temperatures and analysis of trends in the extreme temperature inices. *J. Geophys. Res. Atmos.* **2013**, *118*, 9708–9720.
38. Zhang, X.; Yan, X. Temporal change of climate zones in china in the context of climate warming. *Theor. Appl. Climatol.* **2013**, *115*, 167–175.
39. Zhang, X.; Hegerl, G.; Zwiers, F.W.; Kenyon, J. Avoiding inhomogeneity in percentile-based indices of temperature extremes. *J. Clim.* **2005**, *18*, 1641–1651.
40. Frich, P.; Alexander, L.V.; Della-Marta, P.; Gleason, B.; Haylock, M.; Klein Tank, A.M.G.; Peterson, T. Observed coherent changes in climatic extremes during the second half of the twentieth century. *Clim. Res.* **2002**, *19*, 193–212.
41. Wei, F. *Modern Diagnosis and Prediction of Climate Statistics*, 2nd ed.; Chinse Meteorological Press: Beijing, China, 2007.
42. Kendall, M.G. *Rank Correlation Methods*, 4th ed.; Charles Griffin: London, UK, 1975.
43. Mann, H.B. Non-parametric test against trend. *Econometrica* **1945**, *13*, 245–259.
44. Gocic, M.; Trajkovic, S. Analysis of changes in meteorological variables using mann-kendall and Sen's slope estimator statistical tests in Serbia. *Glob. Planet. Chang.* **2013**, *100*, 172–182.
45. Kadioğlu, M. Trends in surface air temperature data over Turkey. *Int. J. Climatol.* **1997**, *17*, 511–520.
46. Pingale, S.M.; Khare, D.; Jat, M.K.; Adamowski, J. Spatial and temporal trends of mean and extreme rainfall and temperature for the 33 urban centers of the arid and semi-arid state of Rajasthan, India. *Atmos. Res.* **2014**, *138*, 73–90.

47. Sayemuzzaman, M.; Jha, M.K. Seasonal and annual precipitation time series trend analysis in north Carolina, United States. *Atmos. Res.* **2014**, *137*, 183–194.
48. Yue, S.; Pilon, P.; Cavadias, G. Power of the Mann–Kendall and Spearman's rho tests for detecting monotonic trends in hydrological series. *J. Hydrol.* **2002**, *259*, 254–271.
49. Kisi, O.; Ay, M. Comparison of Mann-Kendall and innovative trend method for water quality parameters of the Kizilirmak river. *J. Hydrol.* **2014**, *513*, 362–375.
50. Easterling, D.R.; Horton, B.; Jones, P.D.; Peterson, T.C.; Karl, T.R.; Parker, D.E.; Salinger, M.J.; Razuvayev, V.; Plummer, N.; Jamason, P.; *et al.* Maximum and minimum temperature trends for the globe. *Science* **1997**, *277*, 364–367.
51. Karl, T.R.; Knight, R.W.; Gallo, K.P.; Peterson, T.C.; Jones, P.D.; Kukla, G.; Plummer, N.; Razuvayev, V.; Lindseay, J.; Charlson, R.J. A new perspective on recent global warming: Asymmetric trends of daily maximum and minimum temperature. *Bull. Am. Meteorol. Soc.* **1993**, *74*, 1007–1023.
52. Vose, R.S.; Easterling, D.R.; Gleason, B. Maximum and minimum temperature trends for the globe: An update through 2004. *Geophys. Res. Lett.* **2005**, *32*, L23822.
53. Li, Z.; Liu, W.; Zheng, F. Trends of extreme temperature events in Jinghe watershed during 1965–2005. *Sci. Geogr. Sin.***2010**, *30*, 469–474.
54. Zhai, P.; Pan, X. Change in extreme temperature and precipitation over northern China during the second half of the 20th century. *Acta Geogr. Sin.* **2003**, *58*, 1–8.
55. Shi, J.; Cui, L. Characteristics of high impact weather and meteorological disaster in Shanghai, China. *Nat. Hazards***2012**, *60*, 951–969.
56. Ding, Y.; Wang, Z.; Sun, Y. Inter-decadal variation of the summer precipitation in east China and its association with decreasing asian summer monsoon. Part I: Observed evidences. *Int. J. Climatol.* **2008**, *28*, 1139–1161.
57. Huang, R.; Chen, J.; Wang, L.; Lin, Z. Characteristics, processes, and causes of the spatio-temporal variabilities of the east Asian monsoon system. *Adv. Atmos. Sci.* **2012**, *29*, 910–942.

58. Zhang, R. Changes in east Asian summer monsoon and summer rainfall over eastern China during recent decades.*Sci. Bull.* **2015**, *60*, 1222–1224.

59. Wang, L.; Chen, W. How well do existing indices measure the strength of the east Asian winter monsoon? *Adv. Atmos. Sci.* **2010**, *27*, 855–870.

60. Wang, L.; Huang, R.; Gu, L.; Chen, W.; Kang, L. Interdecadal variations of the east Asian winter monsoon and their association with quasi-stationary planetary wave activity. *J. Clim.* **2009**, *22*, 4860–4872.

61. Qi, L.; Wang, Y. Changes in the observed trends in extreme temperatures over China around 1990. *J. Clim.* **2012**, *25*, 5208–5222.

62. Li, J.; Dong, W.; Yan, Z. Changes of climate extremes of temperature and precipitation in summer in eastern China associated with changes in atmospheric circulation in east Asia during 1960–2008. *Chin. Sci. Bull.* **2012**, *57*, 1856–1861.

63. Ding, T.; Qian, W.; Yan, Z. Changes in hot days and heat waves in china during 1961–2007. *Int. J. Climatol.* **2010**, *30*, 1452–1462.

64. Yao, Y.; Luo, Y.; Huang, J.-B. Evaluation and projection of temperature extremes over China based on CMIP5 model.*Adv. Clim. Chang. Res.* **2012**, *3*, 179–185.

65. Fang, S.B.; Cammarano, D.; Zhou, G.S.; Tan, K.Y.; Ren, S.X. Effects of increased day and night temperature with supplemental infrared heating on winter wheat growth in north China. *Eur. J. Agron.* **2015**, *64*, 67–77.

66. Fang, S.B.; Su, H.; Liu, W.; Tan, K.; Ren, S. Infrared warming reduced winter wheat yields and some physiological parameters, which were mitigated by irrigation and worsened by delayed sowing. *PLoS ONE* **2013**, *8*, E67518.

67. Richardson, A.D.; Keenan, T.F.; Migliavacca, M.; Ryu, Y.; Sonnentag, O.; Toomey, M. Climate change, phenology, and phenological control of vegetation feedbacks to the climate system. *Agric. For. Meteorol.* **2013**, *169*, 156–173.

2

CHAPTER

EXTENSION AGROMETEOROLOGY AS THE ANSWER TO STAKEHOLDER REALITIES: RESPONSE FARMING AND THE CONSEQUENCES OF CLIMATE CHANGE

Kees Stigter [1,2,*], Yunita T. Winarto [3], Emmanuel Ofori [4], Gugulethu Zuma-Netshiukhwi [5], Durton Nanja [6] and Sue Walker [2]

[1] *Agromet Vision-Netherlands, Indonesia, Ghana, South Africa, Zambia, Zimbabwe, Groenestraat 13, Bruchem 5314 AJ, The Netherlands*
[2] *Department of Soil, Crop and Climate Sciences, University of the Free State, P.O. Box 339, Bloemfontein 9301, South Africa*
[3] *Department of Anthropology, Faculty of Social & Political Sciences, Universitas Indonesia, Fl.6/Building H.-Selo Soemardjan Room, Kampus UI, Depok 16424, Indonesia*
[4] *Department of Agricultural Engineering, Kwame Nkrumah University of Science and Technology, Kumasi 03220, Ghana*
[5] *Institute for Soil, Climate and Water, Agricultural Research Council, Private Bag x 79, Glen 9360, South Africa*
[6] *Zambia Meteorological Department, P.O. Box 60004, Livingstone 10101, Zambia*

ABSTRACT

Extension agrometeorology is applied in agrometeorological extension work to advice and serve farmers. In agrometeorology, response farming has been developed decades ago. Climate change complicates response farming, but does not alter it. This paper reports on new operationalization of that response farming in new educational commitments in agroclimatology. It is explained how "Science Field Shops" are an example in Indonesia. This was based on a thorough analysis of what climate change means for farmers in Asia. For Africa, we report on eying the training of agrometeorological extension trainers ("product intermediaries") in West Africa, based on a thorough analysis of what climate change means for farmers in Africa. We also compare experience with reaching farmers in South Africa and farmer communities in Zambia, as new forms of supporting response farming, all under conditions of a changing climate. The paper, for the first time, connects results from four different programs the senior author is taking part in. There is first and foremost the need for training material to make it possible for the product intermediaries to participate in training extension intermediaries. This should, particularly, bring new knowledge to farmers. With what is presently available and with new approaches, climate extension should be developed and tested with farmers in ways that improve farmer preparedness and decision making.

Keywords: changing climate; response farming; extension agrometeorology; Indonesia; Ghana; South Africa; Zambia

1. INTRODUCTION

Climate change has highly complicated the livelihood of farmers in the developing world and will continue to do so as the most recent work of the Climate Change And Food Security (CCAFS) program shows qualitatively and, where possible, quantitatively [1,2]. In some cases there are indications where and why it will go wrong [3,4]. However, as long as the coupling of General Circulation Models (GCMs) based climate models and crop impact models can, for quite some time to come, absolutely not be done with the necessary accuracy for decision making on adaption to climate change [4,5], neither climate modeling nor climate prediction can be of much operational assistance to farmers in most of the world [6]. In detail:

"In assessing impacts (on crops) and any required adaptation options to abate them, future projections of climate and agricultural systems play an important role. Nonetheless, future outlooks of agricultural production and food security are contingent on the skill of GCMs in reproducing seasonal rainfall and temperatures. Thus, accurate climate change projections are important for developing appropriate and effective adaptation strategies (……). In improving projections, enhancing our understanding of important

modes of variability, (......), the responses of plants to (new) environmental factors are key steps to reducing the uncertainties that can potentially constrain adaptation". [4]

Recently [7] pictured extension agrometeorology as a contribution to sustainable development and defined it as agrometeorology that attends to (i) local suffering from weather and climate and persistent ways to diminish it, and (ii) windows of opportunity that (micro)climate offers "on farm". They argued that this has to be done in a participatory approach. This should lead to the establishment of agrometeorological services by and with farmers. It means first and foremost field work with farmers (see also [8]). We define agro-meteorological/-climate services as "All agrometeorological and agroclimatological knowledge and information (including suitable indigenous technologies) that can be directly applied to try to improve and/or protect the livelihood of farmers" (e.g., [8,9]). It was concluded that extension agrometeorology should join hands with other extension fields to mobilize production and protection forces in a multi-functional agricultural production (see also [10]) where trees outside forests (see also [9,11,12]) would play an important role [7].

Now we have the advantage that, in agrometeorology, response farming has been developed decades ago. The principle was widened in [13] and also further developed into a Farm Adaptive Decision Optimization, in [14], widening its fields of operation but keeping the village farmers as focus [15]. In an evaluation, [15] reviewed that response farming is a method of identifying and quantifying, statistically or otherwise, seasonal rainfall variability and (un)predictability and related risks, addressing these risks at the farm level. Originally, response farming was limited to rainfall events, but coping with weather and climate (and often soil) disasters, as well as using windows of weather and climate (and often soil) opportunities, are other forms of responding to weather and climate (and often soil) realities. Services such as advices on design rules on above and below ground microclimate management or manipulation, leading to appreciable microclimatic improvement: shading, wind protection, mulching, other surface modification, drying, storage, frost protection, *etc.*, also belong to such "response farming" agrometeorological services ([15], see also [8]).

Climate change complicates, but does not alter this [13]. Below we report on new operationalizations of that response farming in new educational commitments in agrometeorology. Some were carried out with farmers in Indonesia since 2008 in the new form of "Science Field Shops" [16,17,18,19]. This was done in Gunungkidul (Yogyakarta) and Indramayu (West Java). It was based on a thorough analysis of what climate change means for farmers in Asia [16,20,21]. For Africa, we report on eying the training of agrometeorological extension trainers ("product intermediaries") in West Africa in a new way. This is based on a thorough analysis of what climate change means for farmers in Africa [22,23,24] and on very recent experience with reaching farmers in South Africa [25,26] and farmer communities in Zambia [27]. These are new forms of supporting response farming. It is important to indicate here that in the above mentioned literature [16,22] we consider that as to climate change, three issues have serious consequences for farmers: (i) global warming; (ii) (increasing) climate variability; and (iii) occurrence of more (and possibly more severe) extreme meteorological/climatological events. Response farming has to do with these three issues [16,22]. Due to the fundamentally different ways weather and climate are approached for knowing their future, we consistently talk in this paper about weather "forecasting" but climate "prediction" (see also [25]).

2. INFORMATION, ADVISORIES, SERVICES

Let us first define weather/climate "information", "advisories", and "services". In our view, information in meteorology/climatology is passive in the sense that there is no indication coming with that information on how to use it. "Raw" weather forecasts and climate predictions are examples. Information with recommendations on how to use it, or information otherwise made more client-friendly for solving specific problems, may be called advisories. However, dialogs between senders and receivers are still not necessarily involved. In the USA, an advisory is only an official announcement or warning, while, for example, in India it is a recommendation that is, however, neither compulsory nor enforceable. Establishing such advisories with farmers in their fields should be called services. In that case we have dialogs between farmers and extension. In that case farmers are assisted to carry out the advisories (establish the services) in their fields, jointly, with extension intermediaries and/or farmer facilitators (selected by the farmers from among themselves) who were shown and trained how to do so.

In our fields of work, with small scale and resource-poor farmers, agrometeorological and agroclimatological services (climate services for agriculture) should be established in farmers' fields, with well-trained extension intermediaries and/or (with the upscaling to communities in mind) sufficiently knowledgeable farmer facilitators [16]. These services are then agrometeorological advisories, accepted in their most client-friendly form by these farmers, and molded by them into what could be actually applied in their fields. The services will that way also be jointly validated by farmers and extension. The same applies to other fields of applied agricultural sciences.

From the above, it follows that data and information are useful building stones only if used to build advisories and, preferably, services that can be established in farmers' fields. We report below only on the roads to such results.

3. DEVELOPMENTS IN INDONESIA

3.1. Introduction

From their respective experiences in collaborative ethnography in Asia and participatory agroclimatology in Africa, Winarto and Stigter in 2008 jointly established field work in agrometeorological learning in Indonesia [28]. In this set-up it was proposed to a small group of participating farmers to (i) measure rainfall in their own plots on a daily basis (Figure 1); (ii) make agro-ecosystem observations in those plots, including on soil conditions, water management, pests and progress in crop growth, where necessary also on a daily basis; (iii) discuss these matters with each other and participating scientists and students; and (iv) translate their learning experience in understanding the consequences of climate change for their farming initiatives [29].

Figure 1. Unusual weather patterns helped farmers see the relationship between daily rainfall quantities (as measured by the farmer pictured), soil moisture and the growth of their crops in Indramayu, Indonesia. Photo by Yunita Winarto [29].

These were farmers that carried out rainfed monocropping and multiple cropping. It should be realized that only rainfall is so variable in time and space that it has to be measured in all plots by all participating farmers. This applies anywhere in Africa, Asia and Latin America. All other macroclimate parameters can be taken from the national grid, where necessary; all microclimate parameters can only be well measured by automatic weather stations *in situ*. This is done here and there in industrialized countries in relation to pests and related diseases but nowhere in developing countries [28].

These farmers in Gunungkidul had taken part in a Climate Field School of limited duration but were this way familiar with some aspects of management and manipulation of water in their fields [22,23]. We introduced, from the beginning, sessions in which several scientists of Gadjah Mada University, Yogyakarta, where we were based for this work, were available to answer questions and discuss problems farmers had. In the course of time, these meetings of farmers with applied scientists were called "Science Field Shops" [16]. Such real dialogues with farmers became the way to assess and carry out a response farming to climate change, based on the agrometeorological learning process initiated [15].

In 2009 we were relocated to the Universitas Indonesia in Depok (near Jakarta), on whose behalf we started a set-up of Science Field Shops in Indramayu (north coast of West Java), using our Yogyakarta experience. Presently we work there with about 50 farmers that make the above mentioned measurements/observations and discuss them monthly with each other and with students led by at least one scientist. There are three kinds of (mainly) rice farming carried out here: irrigated, half irrigated (because water only reaches them part of the growing seasons) and rainfed, also pumping groundwater to save a crop where necessary. From the beginning, farmers and applied scientists have jointly tried to relate measurements and observations to rice yields and to yield comparisons between growing seasons, between the same growing seasons in different years and between farmers/fields.

In that way we catch consequences of climate change (as monitored) and consequences of management (sowing, transplanting, water, fighting pests, using fertilizers *etc.*) the way they can be understood with farmers' agrometeorological learning [22]. Also here farmers can ask any agricultural production related question that we then try to answer and also discuss with them in dialogues during Science Field Shops [15]. This way we get answers that apply to stakeholder realities. Farmers now do not feel abandoned, as they largely are by the government extension services, including the pest and disease observers from the Ministry of Agriculture.

3.2. Climate Predictions

The way in which we presently bring the NOAA monthly "ensemble" climate prediction information to the participating farmers in Indramayu, is a more client friendly information that may be called an agrometeorological advisory (see Section 2). We try to derive the most likely start of the main rainy season and also indicate whether the dry season may be expected to be normal. However, our advisory, given as a scenario for the seasons concerned, remains a qualitative one, based on the raw climate prediction information that we receive from NOAA. For example three recent advisories to our farmers (early January, early February and early March, 2013) were:

The seasonal scenario for West Java: "a normal reasonably rainy wet season and a normal start of the dry season that may be expected to be itself normal as well". and

The seasonal scenario should therefore read: "till April the season is expected to remain normal but could go into the wetter end of the normal range". and

The seasonal scenario prediction for the farmers then becomes: "The rainy season will continue as normal but might have rains in March/April at the lower end of what is near normal and it may have an early cessation. The dry season may be expected to be dry indeed with possibly a somewhat earlier start".

The reality from the Indramayu farmers daily measurements was (for a rainy season that had only started by mid-November) a month of February that had appreciably less rainfall than January and March. So from that point of view, the scenario given in January was helpful but that given in February was suggesting the wrong trend within the rainy season, like that in March. But of course these trends remained within the "normal range" of the rainy season as a whole, after a late start. After two years we will ask the farmers through questionnaires about their experiences with these simple seasonal climate scenarios (see [30] on our approach). Dialogues will then help to improve this arrangement.

Of course these messages are translated into Bahasa Indonesia before they are sent via SMS to the farmers concerned. Now when do such advisories become a climate service for agriculture? When such an advisory is discussed with these farmers, and when they can ask questions related to understanding of and acting upon these seasonal scenario advisories, the agrometeorological advisory on climate prediction has got the form of an agrometeorological service. Dialogues will then help to improve this arrangement (Section 2).

3.3. Other Agrometeorological Services (Climate Services for Agriculture)

Part II of [9] has examples of other agrometeorological services and how they were established on-farm (see also [8]). Such services can never be established as a routine without well-trained extension intermediaries and/or farmer facilitators (selected by the farmers from among themselves). What the scientists of the examples in [8,9] actually do is linking science-based (agro)climatological products and problem solving in the agricultural production environment for various categories of farmers and extension intermediaries. This particularly contributes to increasing the useful operational applicability of weather science, climate science and various fields of agricultural sciences for extension agrometeorology. This must be seen as a contribution to science itself [26].

4. THREE THEOREMS OF AGROMETEOROLOGICAL EXTENSION TO BUILD NEW RESPONSE FARMING

We have not long ago reviewed our work on extension agrometeorology [7] as well as on agrometeorological extension [16]. We illustrated how the recent experience in Indonesia very well fits the accepted extension theories and practices [16]. For this paper we review this extension experience in the following essential theorems, not worded this way before. A theorem is a general proposition that is not self-evident but proven by a chain of reasoning. Some of the experience from Africa was shown in [8,9]. The African examples highlighted below in Section 5 may be considered to confirm the sense of this approach.

Extension training should start at the institutions that deliver the agrometeorological and agroclimatological products on which advisories and services should be based.

Reasoning: the weather forecasts, climate predictions, early warnings for agriculture (including those for pest outbreaks), water management issues, microclimate management and manipulations issues, and all other science based information products come from scientific institutes and weather/environmental services. Therefore, that is where extension training should start. In the Climate Field Schools organized in Indonesia, they forgot to thoroughly train these trainers first (e.g., [29,31]). Only recently in some places they start to propose to actually do this [22,32,33,34],

The product intermediaries should in the end train extension intermediaries who work closest with farmers, for the latter intermediaries to be able to deliver advisories and where necessary/possible establish services with

farmers in their fields. Science Field Shops can be used to start up such training.

Reasoning: the present situation is that intermediaries, where they still exist and are sufficiently operational, have hardly been trained in the consequences of climate change for farming and farmers. If they participate two to three years in new educational commitments like Science Field Shops, they should be able to largely take the positions of the scientists that can withdraw and only cooperate when there are (new) problems for which they can assist to find (new) solutions [16].

The extension intermediaries that work closest with farmers should deliver advisories and services to farmers and where applicable establish these services with farmers in their fields. Also these intermediaries can first be trained by participation in the Science Field Shops until product intermediaries exist with enough experience and training.

Reasoning: those extension intermediaries that are supposed to work closest with farmers must in the end be trained by product intermediaries. In the time that it will take to train the latter, also the former can be trained by participation in new educational commitments like Science Field Shops [16].

The conclusions from these extension theorems are our starting points here. However, extension intermediaries are almost everywhere much too small in numbers to serve all farmers of an area/country. It is here that the farmer facilitators come in as another agent that initially needs to be trained in Science Field Shops. Only farmer to farmer extension will in the end be able to reach all farmers that want new knowledge and experience under the conditions of a changing climate. In summary, there is a great responsibility with the scientists involved that have to organize themselves in new educational commitments with farmers to set up field work like Science Field Shops. This is a new participatory way to understand farmers' vulnerabilities, problems and questions, handling response farming under the conditions of a changing climate [15,20].

5. DEVELOPMENTS IN AFRICA

5.1. When It Started and Beyond

From the late eighties and (more successfully) in the nineties as well as much more recently, some agrometeorologists in Africa participated in on-farm research and worked with farmers in solving problems, but nowhere was, for various reasons, official extension part of that picture [9]. This also applies to China with agrometeorologists at provincial weather services [9]. With climate change becoming more and more important, these issues became more and more urgent [9,22,23,24].

On paper, we want to bring researchers, practitioners and users in the field of African climate together to share knowledge and accelerate the harnessing of climate science to the service of decision makers and communities in Africa [35]. But enhancing climate sciences output for Africa is only one of the necessities. If we are not able to considerably increase absorption of climate sciences operational results for decision making within African communities, we are again only filling shelves in libraries or storing material in space within "tablets", but we do not get results used in villages anywhere. Also in the Global Framework for Climate Services, hardly any established system for climate services provision for agriculture is available anywhere near African communities; nor are there many user interface platforms and is there much capacity building in place to make absorption of warnings and other science based products possible. This also applies to other fields than agriculture [26]. This has to do with logistics, but also with the low skill that most products still have [9].

We will report here on a few recent attempts to change that situation with various new response farming approaches.

5.2. Ghana

The work in Ghana was contemplated and frame worked some years ago but became more serious in the course of 2012. It aims at what our first theorem of extension is about, that training of extension trainers should start at the institutions that are supposed to generate science-based agrometeorological products. Countries with agricultural production systems as given farming systems, established as suitable for the conditions before climate change set in, must now, under conditions of a changing climate, use knowledge based systems in order to:

a) Adequately address the new decision making and other old and new needs of the farming communities;

b) Develop agro-meteorological/-climate products/services with and for farmers;

c) Successfully develop and implement new suitable support services for agricultural production (extension, policies, institutions, infrastructure, research, education and training) for their decision making.

d) Agro-meteorological/-climatological services must be further explored. They include among others:

   A. Specific weather forecasts for agriculture, applied and evaluated with farmers [36,37], seasonal climate predictions (see Section 3 above), early warning messages and other disaster preparedness attempts, applied and evaluated with farmers [9];

   B. Design of new cropping/farming systems that can face the new requirements of a farmer first paradigm under a changing climate (e.g., [1,2]);

   C. Generating and supporting a rural response to climate change in agrometeorology [28], which will boil down to improving farmer preparedness and decision making;

   D. Testing and re-evaluation of indigenous coping strategies for the conditions of a changing climate (e.g., [26]).

In view of that, there is a need for reference material and sources of information on extension agrometeorology as there isn't any teaching material that links applied science to development of decision making on relief measures for the farming community in the West African sub-region (and the rest of Sub-Saharan Africa, for that matter). The necessity of coping with climate change makes such educational material even more indispensable. The objectives are to develop training material on extension agrometeorology that can be used as reference material, source of information and training guide for product intermediaries. We have started to write that material. Table A1 (in Appendix) gives a draft Table of Contents. The present content of 282 PowerPoint slides has been recently prepared in Ghana (Stigter and Ofori, unpublished) and has been discussed at length with a local team of scientists. Much work has still to be done.

5.3. Zambia

In [27] the set up of a Community Agrometeorological Participatory Extension Service (CAPES) in Mujika Region, Zambia, was reported. In short, since the community agrometeorological participatory engagement is not meant as an academic exercise of only creating a Community Information Dissemination Plan (CIDP), but must address the community problems, the engagement has to be implemented to get the overall benefits. This works through community involvement under the leadership of the CAPES team. At this stage each stakeholder performs specific tasks as expected, following the CIDP. This stage addresses the overall community problems related to improved food security and livelihoods. This is the most active part of CAPES, where farmers implement the strategies. This process continues in the community's life as long as it is necessary as community livelihood cannot change overnight. Implementation is important for addressing and testing the suitability of strategies in addressing climate related community problems [27].

Farmers participate in the dissemination of climatic knowledge using the multi-disciplinary climatic knowledge dissemination modes such as vernacular radio, radio listening clubs, farmer to farmer dissemination, public small and large group meetings, field experiments, field days, *etc.* Farmer participation makes the CAPES special because after receiving climatic training they engage in knowledge dissemination using different dissemination modes, compelling one another to effective participation and utilization of climate knowledge. The ultimate result is that of an improved livelihood [27].

The current smallholder farmers make decisions (such as choice of crop, variety, planting time, planting density) and select crop management options (such as on thinning the crop, division in time for the use of fertilizers, applications of pesticides) that will fit the season. These choices can then be tested using so called "mother-baby field trials", to find their suitability for the ongoing season. Selected or volunteer farmers manage the baby trials at various locations on their farms while the mother field trial is replicated and is managed by the researchers. Opportunities such as field days held at the mother field for sharing the gained knowledge from these trials are accorded at appropriate times. Lessons learnt at such field days improve on farmers' future decisions, given the advisories/services developed [27].

Farmers require training in the use of climate knowledge where after the presentation of the current climate prediction farmers make decisions based on that prediction, which they test during the season. The increased interaction on climatic knowledge between researchers and community is promoted by field and home visits where farmers will receive additional extension service regarding understanding of climate knowledge and timely on-farm corrections. This will enhance future farmer decision making for improved crop productivity [27].

Farmer to farmer dissemination of knowledge is the most widely used form of climatic knowledge dissemination by farmers within the community. It takes place almost everywhere, any time, given the opportunity. As climatic knowledge is shared by their fellow farmers, who are credible, other farmers easily accept and adopt climate knowledge [27].

5.4. South Africa

Since 2008, research was done on the use of operational weather and climate information in farmer decision making by resource poor as well as commercial farmers in the south-western part of the Free State province, also known as the Modder/Riet catchment, South Africa [25]. It was found that most resource poor farmers in the South-Western Free State, originally, were not familiar with agrometeorological products but relied on their experience and traditional knowledge for farming decision-making (see [26]). Most commercial farmers, having more resources, performed better, compared to the resource poor farmers. Many used (some) agrometeorological advisories before and others were easily introduced to (more of) them. We will therefore continue here mainly with the resource poor farmers that need extension most.

The approach of this research was the establishment of study groups in six towns in the area. The analytical results from questionnaires conducted in 2008 and 2012, with partly overlapping farmers, indicated that farmers in the South-Western Free State differ in agricultural practices, interests, needs, experience, and skills. Therefore, we need provisions of tailor-made products for crop production, animal husbandry, agroforestry, and veld management, respectively. The research, especially, exemplified case studies linking science-based products/advisories and problem solving in the agricultural production environment, for various farmers, where applicable, through extension intermediaries. In the case of this study, they were disseminated to the study groups, of which the composition had been proposed by extension intermediaries involved. This study demonstrated that participating (commercial and) resource poor farmers used weather forecasts and climate predictions for agriculture and other science based agrometeorological advisories during most of the study period [25].

The study groups presented a very diverse situation in terms of experiences, challenges, and needs that are related to their farming. For example, agricultural production in some parts of the study area can be optimal under supplementary irrigation where soil requirements are suitable. The consequences of increasing climate variability needed greater emphasis as to farmer's ability to develop on-farm coping strategies and interventions. A successful farmer should understand the local trends in climate change and how agricultural outcomes are influenced. For example, the concept of response farming prepares the farmer to be aware of past and future climatic conditions and of the extent of increasing climate variability and related dangers and interventions to reduce vulnerabilities. Analyzed long-term rainfall data assisted the farmers to select suitable crops for the area for diversification and sequential planting and determining the suitable planting dates and planting densities. Crop models were used to generate and evaluate a series of management practice scenarios. These outputs from the analyzed climate data and crop models were used to develop advisories that were tailor-made for the farmers. The study for example identified potential production areas for vegetables, herbs/spices, food grains, oil seeds, fruits, and other products such as cotton and other fibers. Application of crop models such as Ehler's model and Eco-Crop 2 revealed the thermal and water requirements of these crops which are either fully or partially met in various parts of the catchment [25].

Participation of farmers in a series of monthly innovative workshops created a conducive environment for knowledge exchange and training. It is advisable to conduct on-farm visits prior to the workshop/meeting days for proper preparations. This was to ensure that the monthly workshops were well-planned, productive, and informative events. The manner in which this study was structured enhanced a bottom-up approach since it allowed participative approaches in close contact with the study groups, improving farming development by closing the gaps existing between developers, suppliers and users of agrometeorological knowledge and understanding. The best outcome of this study was that farmers learnt new things and shared their information and experience. This farmer to farmer extension should be recommended for reaching the highest number of farmers in an area. The remaining challenge for agrometeorological advisories/services providers is to supply reliable and skilled forecasts/predictions and other science based information for agriculture through dissemination methods that suit the farmer [25].

6. DISCUSSION, COMPARING THE VARIOUS RESULTS

Implementation of the engagement of farmers with the CAPES team in Zambia, including house and field visits, may be compared to Science Field Shop meetings, both being attempts to jointly address farming problems. In Zambia, the mother-baby field trials give additional experimentation for discussions and comparisons, where in Indonesia all participating rice farmers meet once a month to compare the results of their measurements and observations in their own plots. Both extension attempts consider the undertakings to be long term strategies. In Zambia, there is already more dissemination by using radio and, after the experimental phase, farmers will continue on their own, assisted by local extension; the same applies in Indonesia after the Science Field Shops phase. Both agrometeorological extension strategies indicate that in the end farmer to farmer dissemination of knowledge is the most important upscaling strategy.

As in Zambia there is community involvement from the beginning, upscaling starts earlier and makes use of the community structures. In Indonesia the learning started one scale below this and the farmer to farmer extension is needed to involve the communities in which the farmers live. The participating farmers in Indramayu are scattered over a rather large area, those in Zambia are from one large community of several villages near to each other. The latter should be considered an advantage at the first upscaling.

The study groups established in South Africa, and in the course of time the monthly workshops with these farmers, prepared by visits prior to these meetings, may also be compared with the Science Field Shops. It is clear that in South Africa, Zambia, and Indonesia, an intensive long-term contact between farmers and scientists involved was the first prerequisite of any success. Extension is nowhere sufficiently trained and active to play a role worth mentioning, and where they participated, they were part of the trainees more than of the trainers. However, Zambia's CAPES could be of subsequent importance in South Africa because "study groups" could in due course be replaced by "communities" as core of the extension methodology.

Once introduced to science-based approaches and products, farmers are most eager to understand the differences between local and scientific knowledge and either combine them, or change, mainly into the use of science-based products. The participating farmers were most varied in South Africa and least so in Indonesia, due to the cropping and farming systems involved. What everywhere is still missing is, first qualitatively and then quantitatively, the influence that the use of science-based products has on yields. Indonesia worked on this most closely by attempting comparisons of rice yields between farmers, between annual seasons and between seasons in the same calendar year.

In South Africa, the work was least field based although improvements in the crop situations were important and readily observed. The mother-baby field trials of Zambia are clearly an important way to observe improvements of knowledge inputs. However, it should be realized that the low yields in Africa need protection from the consequences of climate change most, but can relatively easily be improved. The rice yields in Indonesia vary from total regional failure, due to pests, floods, and drought, to between 5 and more than 10 t/ha on average in normal years everywhere.

Farmer to farmer extension is overall emphasized as crucial, while the high degrees of participation and bottom up extension show to be absolute necessities for any successful involvement of scientists. The more frequently and directly scientists (in the early phases) and well trained extension intermediaries (in the not too far away future) are involved in supporting field management under conditions of a changing climate, the more likely improved livelihoods for farmers become.

7. CONCLUSIONS AND RECOMMENDATIONS

Extension agrometeorology related activities of new response farming, like found in Indonesia, Ghana, South Africa, and Zambia, should be continued and scaled up. Initial operational answers to the changing agricultural reality of stakeholders were found in South Africa and Zambia, and are developing in Indonesia. There is, first and foremost, the need for training material (Appendix, Ghana case study, but also in development in the other cases) to make it possible for the product intermediaries to participate in training extension intermediaries delivering agro-meteorological/-climatological advisories/services to farmers under conditions of a changing climate. This should, particularly, bring new knowledge to farmers [16,21,25,27,38,39,40]. However, they should also be warned for the low skills that many of the products still have, and will continue to have for quite some time to come (e.g., [9,23,30,31,40]).

ACKNOWLEDGMENTS

The senior author thanks the mentioned institutions employing his co-authors Winarto, Ofori and Walker, involved in his regular visits to Indonesia (since 2008) and South Africa (since 2006) and in his first visit to Ghana (in 2012), for their support.

REFERENCES

1. Thornton, P. *Recalibrating Food Production in the Developing World: Global Warming Will Change More than just the Climate*; CCAFS Policy Brief No. 6; CGIAR Research Program on Climate

Change, Agriculture and Food Security (CCAFS): Copenhagen, Denmark, 2012.

2. Climate Change, Agriculture and Food Security (CCAFS), *Impacts of Climate Change on the Agricultural and Aquatic Systems and Natural Resources within the CGIAR's Mandate*; Thornton, P., Cramer, L., Eds.; CCAFS Working Paper 23; CGIAR Research Program on CCAFS: Copenhagen, Denmark, 2012.
3. Gourdji, H.M.; Sibley, A.M.; Lobell, D.B. Global crop exposure to critical high temperatures in the reproductive period: Historical trends and future projections. *Environ. Res. Lett.* **2013**.
4. Ramirez-Villegas, J.; Challinor, A.J.; Thornton, P.K.; Jarvis, A. Implications of regional improvement in global climate models for agricultural impact research. *Environ. Res. Lett.* **2013**.
5. Van Oldenborgh, G.J.; Doblas Reyes, F.J.; Drijfhout, S.S.; Hawkins, E. Reliability of regional climate model trends.*Environ. Res. Lett.* **2013**.
6. Anonymous. Extreme weather. Better models are needed before exceptional events can be reliably linked to global warming. *Nature* **2012**, *489*, 335–336.
7. Stigter, K.; Winarto, Y.T. Extension agrometeorology as a contribution to sustainable agriculture. *New Clues Sci.* **2012**,*2*, 59–63.
8. Stigter, C.J. *Agrometeorological Services: Reaching all Farmers with Operational Information Products in New Educational Commitments*; CAgM Report 104; World Meteorological Organization (WMO): Geneva, Switzerland, 2011.
9. Stigter, K. *Applied Agrometeorology*; Springer: Heidelberg/Berlin, Germany/New York, NY, USA, 2010.
10. Stigter, K.; Walker, S.; Das, H.P.; Huda, S.; Zheng, D.; Liu, J.; Li, C.; Dominguez-Hurtado, I.M.; Mohammed, A.E.; Abdalla, A.T.; *et al.* Meeting Farmers' Needs for Agrometeorological Services: An Overview and Case Studies. In*Addressing the Livelihood Crisis of Farmers: Weather and Climate Services*; Available online: http://www.cmp.vcl.cu/uploads/documentos/fc9a911af18abaa5d0512cf015a3217ab3f59544.pdf (accessed on 15 March 2012).
11. Stigter, K. Agroforestry in coping with meteorological and climatological risks. *The Overstory* **2011**. the Overstory 233.

12. Stigter, K.; Ofori, E.; Kyei-Baffour, N.; Walker, S. Microclimate Management and Manipulation in Applied Agroforestry. *The Overstory* **2011**. the Overstory 240.
13. Stigter, K. Agrometeorological services under a changing climate: Old wine in new bags. *WMO-Bull.* **2008**, *57*, 114–117.
14. Gommes, R.; Acunzo, M.; Baas, S.; Bernardi, M.; Jost, S.; Mukhala, E.; Ramasamy, S. Communication Approaches in Applied Agrometeorology. In *Applied Agrometeorology*; Stigter, K., Ed.; Springer: Heidelberg/Berlin, Germany/New York, NY, USA, 2010; pp. 263–286.
15. Stigter, C.(K.)J.; Winarto, Y.T. Coping with Climate Change: An Active Agrometeorological Learning Approach to Response Farming. In *Harnessing and Using Climate Information for Decision Making: An In-Depth Look at the Agriculture Sector*; Asian Pacific Economic Council (APEC): Busan, Korea, 2012.
16. Stigter, C.(K.)J.; Winarto, Y.T. Science Field Shops in Indonesia. A start of improved agricultural extension that fits a rural response to climate change. *J. Agric. Sci. Appl.* **2013**, *2*, 112–123.
17. Winarto, Y.T.; Stigter, K.; Anantasari, E.; Prahara, H.; Kristyanto. Collaborating on establishing an agrometeorological learning situation among farmers in Java. *Anthropol. Forum* **2011**, *21*, 175–197.
18. Winarto, Y.T.; Stigter, K.; Dwisatrio, B.; Nurhaga, M.; Bowolaksono, A. Agrometeorological learning strengthening farmers' knowledge to cope better with climate change. *J. Southeast Asian Stud.* **2013**. in print.
19. Winarto, T.; Stigter, K. Science field shops to reduce climate vulnerabilities: An inter- and trans-disciplinary educational commitment. *Collab. Anthropol. (University of Nebraska, USA)* **2013**. in press.
20. Stigter, C.(K.)J.; Winarto, Y.T. Considerations of climate and society in Asia (I): What climate change means for farmers in Asia. *Earthzine* **2012**.
21. Stigter, C.(K.)J.; Winarto, Y.T. Considerations of climate and society in Asia (II): Our work with farmers in Indonesia.*Earthzine* **2012**.
22. Stigter, C.(K.)J.; Ofori, E. What climate change means for farmers in Africa. Part One: Introductional matters and consequences of global

warming for African farmers. *Afric. J. Food Agric. Nutr. Dev.* **2013**. in press.

23. Stigter, C.(K.)J.; Ofori, E. What climate change means for farmers in Africa. Part Two: Increasing climate variability and a response approach for African farmers. *Afric. J. Food Agric. Nutr. Dev.* **2013**. submitted.
24. Stigter, C.(K.)J.; Ofori, E. What climate change means for farmers in Africa. Part Three: Climate extremes and society's responses, including mitigation attempts as part of preparedness of African farmers. *Afric. J. Food Agric. Nutr. Dev.* **2013**. submitted.
25. Zuma-Netshiukhwi, G.N. The Use of Operational Weather and Climate Information in Farmer Decision Making, Exemplified for the South-Western Free State, South Africa. Unpublished Ph.D. Thesis, University of the Free State, Bloemfontein, South Africa, 2013.
26. Zuma-Netshiukhwi, G.; Stigter, K.; Walker, S. Use of traditional weather/climate knowledge by farmers in the South-western Free State of South Africa: Agrometeorological learning by scientists. 2013. submitted.
27. Nanja, D.; Walker, S. *Handbook for Community Agrometeorological Participatory Extension Service (CAPES)*; Meteorological Department-University of the Free State: Livingstone/Bloemfontein, South Africa/Zambia/, 2011.
28. Winarto, Y.T.; Stigter, K. *Agrometeorological Learning: Coping Better with Climate Change*; LAP LAMBERT Academic Publishing GmbH & Co. KG: Saarbrucken, Germany, 2011.
29. Winarto, Y.T.; Stigter, K.; Anantasari, E.; Hidayah, S.N. Climate field schools in Indonesia: Improving "response farming" to climate change. *LEISA Mag.* **2008**, *24*, 16–18.
30. Stigter, C.(K.)J.; Winarto, Y.T. APCC Guidance Questions and Our Replies Related to Handling Seasonal Climate Predictions with Farmers. Available online: http://www.agrometeorology.org/topics/accounts-of-operational-agrometeorology/apccguidance-questions-and-our-replies-related-tohandling-seasonal-climate-predictions-with-farmers/(accessed on 15 May 2013).

31. Indonesian Agency for Meteorology, Climatology and Geophysics (BMKG), *Climate Field Schools at District Level in Eastern Indonesia*; BMKG: Jakarta, Indonesia, 2011.
32. Al-Amin, N.K.N. Personal Communication. Khartoum, Sudan, 2012.
33. Rivero Vega, R.E. Personal Communication. Camaguey, Cuba, 2013.
34. Zuma-Netshiukhwi, G. Personal Communication, Glen, South Africa, 2013.
35. ClimDev-Africa. African Climate Science Frontiers: Addressing Priority Research Gaps to Inform Adaptation Decision-Making in Africa. In To be presented at 2013 Africa Climate Conference, Arusha, Tanzania, 15–18 October 2013. in press.
36. Murthy, R.K. Field Exercise on "Murthy's Daily Weather-Agriculture". *"Hands on" Training for Response Farming. Reactions to Calls for Information Collection and Exchange*, Available online: http://www.agrometeorology.org/topics/hands-on-training-for-response-farming (accessed on 15 May 2013).
37. Murthy, R.K. *Report on the Project Entitled "WMO-ANGRAU-DST Sponsored Roving Seminars on Weather, Climate and Farmers"*, Available online: http://www.agrometeorology.org/topics/needs-for-agrometeorological-solutions-to-farming-problems/report-on-the-project-entitled-201cwmo-angrau-dst-sponsored-rovingseminars-on-weather-climate-and-farmers (accessed on 15 May 2013).
38. Stigter, C.J. Agrometeorology from science to extension: Assessment of needs and provision of services. *Agric. Ecosyst. Environ.* **2008**, *126*, 153–157.
39. Stigter, C.J.; Oteng'i, S.B.; Oluwasemire, K.O.; Al-Amin, N.K.; Kinama, J.M.; Onyewotu, L.O. Recent answers to farmland degradation illustrated by case studies from African farming systems. *Ann. Arid Zone* **2005**, *44*, 255–276.
40. Hansen, J.W.; Mason, S.J.; Sun, L.; Tall, A. Review of seasonal climate forecasting for agriculture in sub-saharan Africa. *Exp. Agric.* **2011**, *47*, 205–240.

3

CHAPTER

REMOTE SENSING BASED CROP COEFFICIENTS FOR WATER MANAGEMENT IN AGRICULTURE

Jorge Rocha[1], Raquel Melo[1], Cristina Henriques[1] and António Perdigão[2]

[1] *CEG, Institute of Geography and Territorial Planning, University of Lisbon,, Portugal*

[2] *DGADR, Direcção Geral de Agricultura e Desenvolvimento Rural, Lisbon,, Portugal*

1. INTRODUCTION

The climatic factors and their variability, both spatial and temporal, linked to precipitation decreasing and irregular distribution, due to climatic changes, have been gathering a higher weight in the definition of water management policies. These policies have important implications on agriculture. Using new technologies that allow a better use of water requires institutional changes in major areas. The first point is the need for base information with an adequate spatial and temporal resolution. The work we have done includes itself in the water efficient and sustained use, allowing the improvement of irrigation systems and it's the result of a jointly effort of several teams based on an international project.

PLEIADES (Participatory multi-Level EO-assisted tools for Irrigation water management and Agricultural Decision-Support) is a research and technological development project co-funded by the European Commission's Sixth Framework Programme within its Sustainable Development, Global Change and Ecosystems Priority. The project responds to the identified need for targeted research in the area of Integrated management strategies and mitigation technologies, topic Water in Agriculture: new systems and technologies for irrigation and drainage. A set of pilot Case Studies represents a sample of the wide range of conditions found in the European and Southern Mediterranean and in the Americas, covering Portugal, Spain, Italy, Greece, Turkey, Morocco, Mexico, Peru, and Brazil.

The Portuguese working area was the Caia irrigation area, a subsystem of Guadiana basin, located in the southeast of Portugal, near the border with Spain. PLEIADES was expected to generate new knowledge on the functioning and performance of these pilot areas. This in turn aimed at providing the knowledge and information base for decision makers at all levels on agricultural water needs and consumption. It also set out to provide the basis for assessing the benefits and threats potentially brought about by new technologies to all actors in changing environments. The project was also expected to generate new tools for irrigation water management, combining innovative sensor technology with flexible easy-to-use Decision-Support Systems for adaptive management. These tools were designed to help farmers to control water more efficiently and improve the environmental and economic performance of their irrigation systems.

In our work, we address the efficient and sustainable use of water for food production in water-scarce environments. We consider the economic, environmental, technical, social, and political dimensions through a synergy of leading-edge technologies and participatory approaches. These technologies provide easy access to information for all stakeholders while active participation will be effected by spatial information and innovative networking tools. Our aim is to improve the performance of irrigation schemes by means of a range of measures. Major technical innovation is made possible by the comprehensive space-time coverage of *EO* data and the interactive networking/connecting capabilities of Information and Communication Technologies (*ICT*).

The project was designed to assess and demonstrate in an operational perspective how the integration of Earth observation (*EO*) techniques in routine Irrigation Advisory Services (*IAS*) can improve the efficiency in the use of water for irrigation. The use of leading-edge Information and Communication Technology (*ICT*) tools in the generation and distribution of information makes the *EO* easily available to *IAS* and the farmers. The project WebGIS (www.pleiades.es) was one of the central outcomes of our project. Its key feature is the operational generation of irrigation scheduling information products from a virtual constellation of *EO* satellites and their delivery to farmers in near-real-time using leading-edge on-line analysis and visualization tools. It is supported by a methodology package to derive crop coefficients and further advanced parameters from *EO* satellite images in an operational processing chain.

The overall goal is to improve and optimise irrigation and drainage systems by means of new technologies. In this context, New Technologies (*NT*) include Earth observation, Geographical Information Systems, Information and Communication Technologies, and Decision-Support systems. In order to achieve this overall goal 3 specific objectives have been set:

1. In accordance with the identified needs of stakeholders, multi-level NT-assisted tools will be adapted and developed for farmers and other water managers to optimise their water use at farm, irrigation scheme and at river-basin levels.
2. To conduct trial campaigns in pilot areas with the active participation of users at farm and irrigation scheme level.
3. To evaluate the performances of the NT-assisted tools using an extended evaluation system covering technical, economic, environmental, social and political dimensions and involving stakeholders at all levels over the whole lifetime of the project.

PLEIADES aims to demonstrate that New Technologies (*NT*) can effectively support the optimisation of irrigation schemes and in the long run foster sustainability by providing comprehensive and timely spatial information that supports decisions made at many levels: farms, irrigation schemes and river-basins.

We recognise that improvements will come not only from technical innovations but also from changes in social factors related to water governance, participation and social learning. Thus the NT-assisted tools will be deployed to facilitate technical and social learning enabling farmers to act responsibly by fine tuning their on-farm practices in accordance with the river-basin water status.

This work addresses the efficient and sustainable use of water for food production in water scarce environments. It aimed to improve the performance of irrigation schemes by means of a range of measures that consider the economic, environmental, technical, social, and political dimensions through a synergy of leading-edge technologies and participatory approaches. Major social and technical innovation was made possible by the comprehensive space-time coverage of Earth observation (*EO*) data and the interactive networking/connecting capabilities of Information and Communication Technologies. The system we developed stands mainly over *FAO* normative, about culture water needs and the calculation of cultural coefficient (*Kc*) in a simple way, directly from remote sensing data. For that we simply use radiometric parameters derived from visible and infrared bands. Crop evapotranspiration can be calculated using the crop coefficient (*Kc*) defined as the ratio of total evapotranspiration (*ET*) by reference evapotranspiration (*ET0*).

Earth Observation (*EO*) provides an objective evaluation of crop water demand; this information can be used at different decision levels (from the farmers to the river basin authorities) to promote a more efficient use of water resources in agriculture. A rational management of water resources for irrigation requires information characterized by high temporal and spatial variability, which cannot be monitored with traditional field inspections. *EO* is a mature technology, ready for being transferred to operational applications in agricultural water management. Detailed data on crop development and irrigation needs are timely distributed to final users by means of modern Information and Communication Technologies.

Three main usages of *EO* based products have been conceived:

1. Distribution of personalized information to a range of stakeholders (i.e. landowners, irrigation farmers and their associations) concerning crop and water status;
2. Integration in *GIS* based river-basin water management tool, for distributed water balance calculations.
3. A portfolio of *EO* based products has been set-up, and the methodologies for their retrieval have been defined, starting from past experiences and scientific knowledge available among the partners in the Consortium. *FAO*. methodology has been adopted as the standard procedure for computing crop water requirements from *EO* based products.

Three different levels of *EO* based products are distinguished:

1. Land-use (irrigated vs. non irrigated crops; crop inventory maps);
2. Basic (vegetation cover, Leaf Area Index, Crop Coefficients, potential evapotranspiration, Crop Water Requirements among others);
3. Advanced (reference and actual evapotranspiration, biomass, yield).

The conceptual approach for the derivation of *EO* model of the service is split up into the following steps:

1. Acquisition and analysis of high resolution satellite images in the visible and infrared spectrum;
2. Local agro-meteorological data acquisition (e.g. temperature, humidity, wind speed, sun radiation, rainfall);
3. Field validation through measurements in selected areas;
4. Elaboration of *EO* based products;
5. Data quality check and integration in a dedicated Geographical Information Systems (*GIS*) for irrigation management from field to district and hydrological basin scale;
6. Real-time distribution of personalized irrigation advices on a weekly basis directly to farmers by means of different communication systems (Internet, text and graphical messages by using*GSM/UMTS*).

The validation of the different methodologies for the retrieval of *EO* based products has been an important part of the work carried out within all the pilot areas. Intensive field campaigns carried out simultaneously to satellite acquisitions have produced a large dataset for calibration and validation purposes. Micrometeorological instrumentations have been installed for comparison between field measurements of crop water use and estimates from *EO* processing. New methodologies have been set-up i.e. for improving the estimation of canopy parameters and for calculating reference evapotranspiration from geostationary satellites (of particular relevance in areas with very limited meteorological data).

2. VIRTUAL CONSTELLATION AND SPACE SEGMENT OPERATIONALITY

Farm management requires monitoring of agricultural crops at high spatial resolution and frequent temporal coverage during the entire growing season. The necessary spatial resolution can be provided by the current high-resolution sensors (20-30m pixel size), like TM, ETM+, SPOT, LISS, ASTER, ALI, or in the case of agricultural plots with special spatial requirements by very-high resolution sensors (like Quickbird, Ikonos). However, canopy architecture and biophysical parameters describing the canopy, like leaf area index, fractional ground cover, biomass, evapotranspiration, water stress, evolve continuously during the crop growing season.

A single satellite with a 16 day repeat time (like Landsat) would provide little useful information, considering also that cloud conditions may increase the time period between useable images. The critical requirement of frequency of coverage combined with high spatial resolution has not been satisfied after more than thirty years of Landsat mission launching. Satellite constellations have been proposed for this purpose by some studies, but a practical solution for the near future is not at hand. Our solution is a virtual constellation of EO satellites that corresponds to the needs and user requirements of irrigation scheduling and precision agriculture. Therefore, we have developed a unified procedure to obtain consistent time series of vegetation parameters of interest from this virtual constellation.

2.1. Definition and Operational Aspects

The virtual constellation (*VC*) is defined as a set of *EO* satellites, each of which provides the necessary data to derive *NDVI* and other vegetation parameters at the spatial resolution required for the given application. In our case, for irrigation scheduling a spatial resolution of at least 30 m is imperative to resolve the major part of agricultural fields (plots). An additional key selection criterion is near-nadir observation, such that bi-directional effects are minimized. Landsat is the backbone of this *VC*, because of its excellent operational availability and low cost.

Due to the technical failure in Landsat7 all images taken on/after 31 May 2003 have no useable data in a significant part of each image. The software correction offered by USGS involves spatial degradation, which moves the spatial resolution of ETM+ out of the interest margin of our *VC*, except for areas located directly in the centre of a scene (non-affected area). This sensor failure clearly demonstrates the vulnerability of the operational space segment for this and similar applications.

Currently available alternative platforms are less ideal for operations, since they are either much more expensive (IRS, ALI, Spot), more complicated (Spot, due to changing view angles), and/or not operational (Aster, no rush service). A number of emerging platforms may add more elements to the *VC*. Our experience in PLEIADES has shown that all can be used to complete the TM time series. Ikonos and Spot were successfully tested in the Italian pilot area. The experience of Spot programming for this area showed that on average 5-6 images per month can be obtained (fairly cloud free and with incidence angles less than 15º).

We want to stress here again that the space segment is the most vulnerable part of the entire operational system. For this reason, urgent actions are required to ensure the capability to obtain adequate *EO* images at the adequate coverage frequency and low cost. As a practical near-term solution for the case of cloudiness and/or satellite sensor failure, a contingency scenario was developed to base the PLEIADES operational system on a synergistic combination of *EO* data, field data, and an expert system of local crop coefficient (*Kc*) curves.

These curves have been developed from the synthesis of previous campaigns, specially tailored to the crops and climatology of a given area. In the case of an *EO* data failure (either missing image or clouds), the system would draw on a default list of *Kc* curves (per crop, crop cycle, sowing date) from (in order of priority) field data, the local expert system data base, and the look-up tables recommended by the Food and Agriculture Organization (FAO) [1]. Medium-resolution sensors (like AVHRR or MODIS) are also used to derive support data for this purpose.

2.2. Methodology to Derive Pleiades Parameters from EO Data

All biogeophysical parameter calculation from EO starts with a pre-processing of EO data, composed of three main steps: i) Geometric correction and image re-sampling; ii) Cross-satellite intercalibration; iii) Atmospheric correction.

Semi-automatic procedures have been developed in order to elaborate *Kc* maps from *EO* data in the minimum possible time. Pre-processing requires approximately half of the elaboration time of the entire process. Once georeferenced surface reflectance has been calculated in each pixel, the algorithms for determining *Kc* are quite straightforward.

The elaboration for step (i) is based on consolidated procedures available in each pilot zone. This step does not necessary requires a standardization, and it strictly linked to the topographical mapping standard adopted in each area. The principal recommendation coming from PLEIADES is to adopt procedures that do not alter substantially the radiometric content of data; as such, first-degree relationships should be preferentially used for coordinates transformation and nearest neighbour techniques for pixel resampling.

2.3. Cross-Sensor Intercalibration

When using different sensors from our virtual constellation to generate time series of maps of geobiophysical parameters, a reliable methodology is needed to intercalibrate the observations from different sensors at different observation scales in different platforms. Intercalibration between observations or cross-calibration of sensors aims at developing relationships that allow to translate reflectances and spectral vegetation indices from one sensor to another.

For this purpose, we have performed an observational study, comparing reflectances and *NDVI* from near-synchronous image pairs of ETM+ as the reference sensor and TM, LISS, Aster, Quickbird, and AVHRR. Linear relationships were found for the intercalibration of reflectances and *NDVI* from one sensor to another, for all sensors, provided that some spatial aggregation is performed.

The main source of data dispersion in our linear cross-sensor translation equations is the geolocalization uncertainty inherent in the process of geometric correction. Consequently, spatial aggregation needs always to be performed if (different or the same) sensors are to be used to derive time-series of biogeophysical parameters over heterogeneous areas.

The homogenous zone approach developed here is recommended as an excellent tool for deriving robust new cross-sensor relationships, provided that the selected homogeneous crops cover the full*NDVI* range. The linear cross-sensor relationships derived from one image pair are shown to be valid for the whole season and for all areas with similar vegetation and climate. We recommend repeating the procedure once or twice a year in order to check the temporal stability of the radiometric calibration coefficients.

Although the differences in most cases are small, we maintain different equations for different spatial aggregation sizes. Table 1 gives the coefficients of the resulting equations for the example of 100 m grid cell side length (except for 75 m for Aster), which is the minimum recommended grid size.

TABLE 1. Summary of NDVI cross sensor translation equations for elements in our virtual constellation. Measure of uncertainty is defined as deviation from the mean.

Sensor	Coefficients of linear equation $NDVI - ETM = a \cdot NDVI(sensor) + b$	
	a	b
TM	1.0336	+0.0128
LISS III 1D	1.1672	-0.0454
Quickbird	1.0443±0.008	+0.0191±0.005
Aster L1B	1.1304±0.022	-0.0002±0.019

2.4. Operational Atmospheric Correction

An automatic procedure has been developed for atmospheric correction. This procedure uses parameters extracted directly from the image instead of recurring to external data on synchronous on-site vertical profiles of atmospheric data, which are usually difficult to obtain. It results in a substantial reduction of processing time as compared to approaches based on radiative transfer calculations. It was found that approach (i) for *NVDI Kc* calculation is less sensitive than approach of analytical *Kc* (ii) for the effects related to the atmospheric correction.

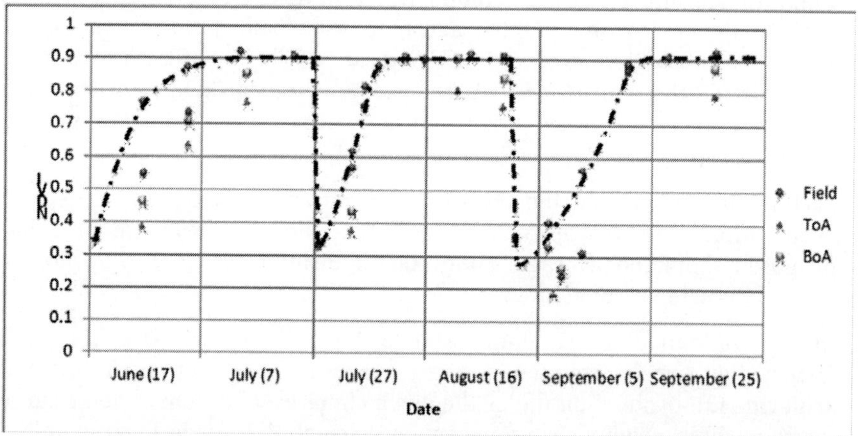

FIGURE 1. Example of the implementation of the atmospheric correction (*AC*) module on the satellite derived NDVI for an alfalfa plot. Top of atmosphere (*ToA*) *NDVI* refers to the *NDVI* without *AC*, whereas of atmosphere (*BoA*) *NDVI* corresponds to *NDVI* with *AC*. The reference field measurements of *NDVI*are shown as solid circles.

A validation of the performance of this module has been performed by using field measurements of spectral reflectance carried out with the spectroradiometer GER 3700 (331 - 2509 nm spectral coverage, 1.5 - 9.5 nm resolution) for two land uses: (i) an alfalfa plot and (ii) a bare soil. Figure 1illustrates the outcome of the AC-module in terms of the *NDVI* values for the alfalfa plot. Field measurements of *NDVI* showed the greatest scattering due to both, heterogeneity of the canopy cover in the alfalfa field and, the small field of view of the spectroradiometer in comparison to Landsat pixel size. Taking into account this variability, we can conclude that atmospherically corrected *NDVI* values are in agreement with in field measured *NDVI*. Similar results are obtained for the bare soil plot.

3. CROP COEFFICIENT AND REFERENCE EVAPOTRANSPIRATION

Crop evapotranspiration can be calculated using the crop coefficient (*Kc*) (1) defined as the ratio of total evapotranspiration (*ET*) by reference evapotranspiration (*ET0*). Combining *Kc* (from field measurements or from satellite images) with *ET0* from agrometeorological station observations allows us to calculate crop evapotranspiration. This coefficient integrates the effect of characteristics that distinguish a typical field crop from the grass reference, which has a constant appearance and a complete ground cover.

Factors that determine the crop coefficients are crop type, climate, soil evaporation and crop growth stages [1, 2]. For this purpose FAO has proposed tabulated average values distinguishing by crops that can be applied knowing its phenology.

In case of annual crops under standard conditions (disease-free, well fertilized, grown in large fields, under optimum soil water conditions and achieving full production under the given climatic conditions), the *Kc* curve for the whole growing season can be calculated considering the initial (*KcINI*), medium (*KcMID*) and end stage (*KcEND*).

$$K_c = \frac{ET}{ET_0} \tag{1}$$

Mainly at the initial and end period, due to lower values of crop cover, soil evaporation has a large effect on *KcINI* and *KcEND* [3, 1, 4]. Therefore, vegetation indices (*VI*) are better related to transpiration of crop than to *Kc* [5] in those periods. This introduces a great variation in *KcINI* and *KcEND* daily values depending on soil water status, i.e. on frequency of wetting by irrigation and rainfall.

The dual crop coefficient approach proposed by [6] splits *Kc* into separate coefficients, one for crop transpiration *Kcb* (basal crop coefficient), and one for soil evaporation (*Ke*). The soil evaporation coefficient, *Ke*, describes the evaporation component of *ET*. When topsoil is wet, after irrigation or rainfall, *Ke* is maximal.

Estimation of *Ke* requires knowledge of soil water balance [1]. Wright [6] introduced the idea of a basal crop coefficient in which the soil evaporation component of *ET* was minimal due to a dry soil surface but adequate soil moisture in the crop root zone was available.

$$ET = \left(K_{cb} + K_e\right) \cdot ET_0 \qquad (2)$$

In PLEIADES we use two approaches to obtain the crop coefficient from satellite imagery: one, directly from *NDVI*, named *Kc - NDVI*, based on the relationship between *NDVI* and the basal crop coefficient, and another, named analytical *Kc*, is based on the direct application of the Penman-Monteith equation. The *NDVI* is the main operational parameter to monitor vegetation status using Earth Observation.

4. CROP COEFFICIENT FROM NDVI AND CANOPY BIOPHYSICS PARAMETERS

Relevant canopy biophysics parameters are green fractional cover, fraction of absorbed photosynthetically active radiation, primary production, Leaf Area Index (*LAI*), basal crop coefficient. All they are involved in canopy evapotranspiration. The ability of *NDVI* to describe canopy biophysics parameter has been shown as follows:

1. *NDVI* is related linearly with green fractional cover [7, 8];
2. *NDVI* is related linearly with the fraction of absorbed photosynthetically active radiation (*fAPAR*) [9];
3. *NDVI* is related with primary production (dry biomass) by means of Light Use Efficiency (*LUE*) models [10, 11, 12]:

$$P = \int_0^t \left(a \cdot NDVI + b\right) \cdot PAR \cdot \varepsilon \cdot W \cdot dt \qquad (3)$$

where *P* is primary production, *PAR* is Photosynthetically Active Radiation, is the efficiency of crop to transform PAR into dry mass, W is a water stress coefficient, and a, b are constants. Using these LUE models we can consider that, under non-water stress, NDVI on plateau stage can be seen as a good estimator of the dry matter accumulation rate, depending on crop and environmental variables [13].

It establishes a relationship between NDVI and crop growth rate (CGR) which agrees with the idea that considers NDVI as an estimator of the canopy photosynthetic power. This way, [14] consider that vegetation index can be legitimately used to provide an estimate of growth rate.

1. NDVI is related exponentially with Leaf Area Index (LAI) [15]. Is well known that NDVI begins to saturate for a value of LAI equal to 3 reaching a plateau for LAI > 3 (Figure 2 a);

2. NDVI is related linearly with the basal crop coefficient K*cb* [16]. This relationship is a relevant basis for the K*c* – NDVI approach.

The facts pointed out in (3) and (4) may appear contradictory (saturation of NDVI for LAI > 3 on one hand and the linear relation of NDVI with K*cb* on the other). This seeming paradox is due to the usual reasoning that relates higher LAI with higher evapotranspiration. This reasoning arises from associating more leaf surface with more transpiration. However, already [17] stated that the evidence seems conclusive that transpiration in most mesophytic crop plants and other mesophytic vegetation well supplied with water increases with leaf area to LAI of about three (Figure 2 b).

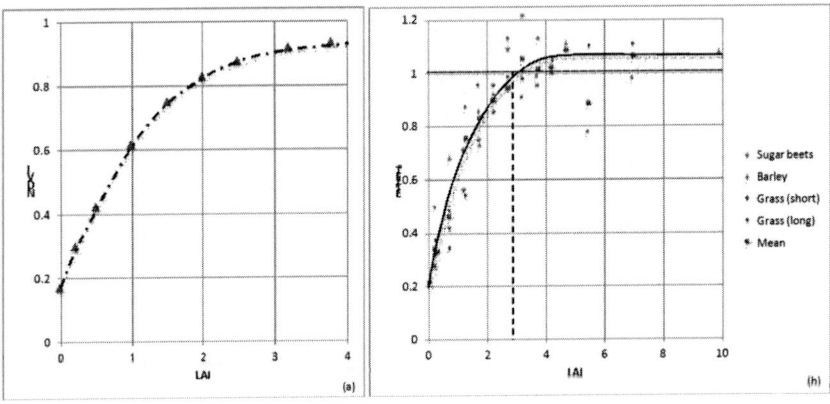

FIGURE 2. Typical NDVI-LAI curve (a), and ratio of actual to potential evapotranspiration (ET/ET*p*) as a function of the LAI (b).

Accounting the LAI saturation in the relation with evapotranspiration has lead to the concept of active LAI (Allen et al., 1998). The active LAI is defined as the index of the leaf area that actively contributes to the surface heat and vapour transfer. It is generally the upper, sunlit portion of a dense canopy. For practical applications, however, the active LAILAIis an ambiguous concept due its dependence on canopy architecture and its interaction with sunlight.

The basal crop coefficient is clearly related with green fractional cover (*fc*) (Figure 3). In fact, the procedure to estimate *Kcb* is based in the knowledge of *fc* [1], despite of ambiguities of the green *fc*concept, mainly in the maturation stage. The relationship between NDVI and CGR for well watered crops is based on the ability of NDVI to estimate fAPAR, introducing this fact in the LUE model. CGR is also related with the transpiration rate.

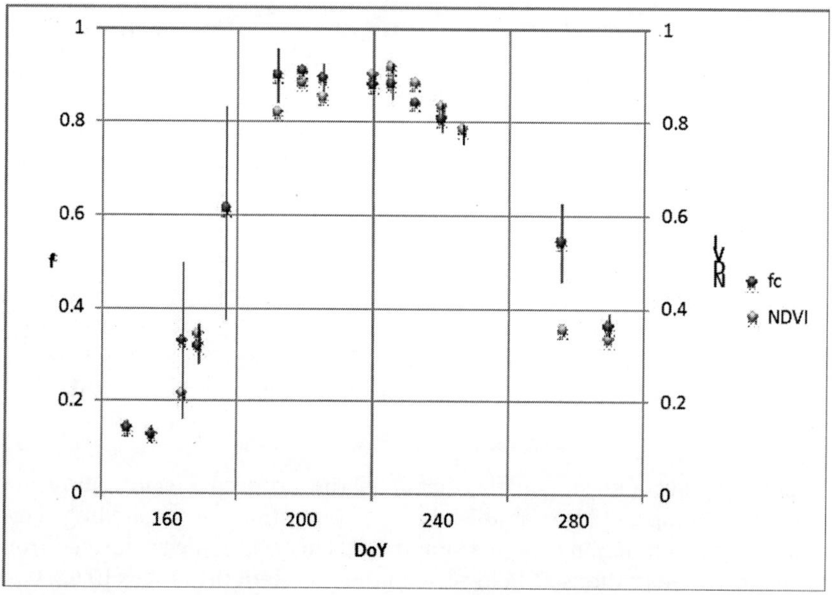

FIGURE 3. Green Fraction Coverand NDVI versus DoY at field scale for corn.

The relationship between NDVI and CGR exhibits a strong dependence on crop and environmental variables (solar radiation, temperature, etc.) [13]. This is due to the different nature of NDVI and CGR. NDVI depends only on canopy characteristics, while CGR and so the transpiration rate, are strongly also dependent on surface and environmental variables. The basal crop coefficient is the ratio of canopy transpiration rate over the reference canopy transpiration rate.

So, the empirical relationship between NDVI and *Kcb* shown by many authors, e.g. [18] and [19], (Figure 4) could be explained by considering NDVI as a measurement of relative CGR. Further research will be need in this subject. Despite limitations due to variability associated with canopy structure, background soil, and calibration uncertainties, NDVI can be used advantageously to estimate crop water requirements [20] in accounting its relationship with K*cb*.

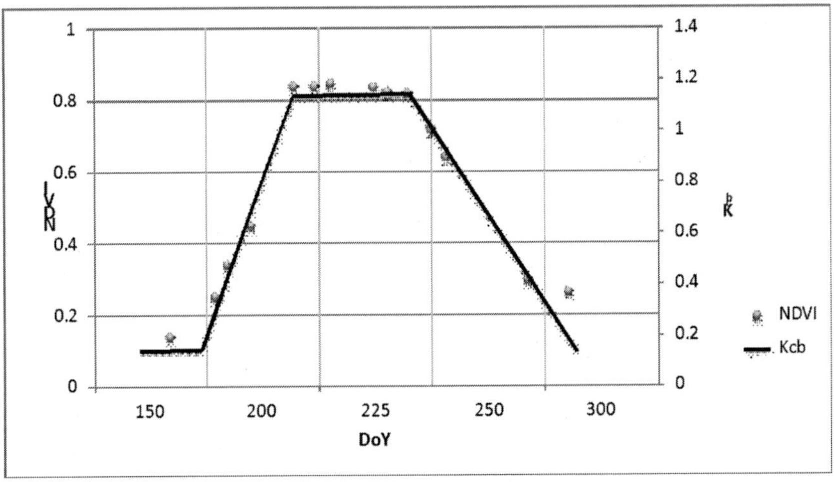

FIGURE 4. Temporal evolution of crop coefficient (K*cb*) and NDVI, in maize. K*cb* is estimated from green plant cover using FAO methodology [19].

Taking into account similarities between the crop coefficient curve and vegetation index, [21] established the potential for modelling crop coefficient as a function of vegetation index. This relation was derived from reflectance observations at field scale in the wavelengths ranges [0.63, 0.69 m] and [0.76, 0.90 m], measured at nadir and two meters above corn. A linear transformation of the NDVI was developed by equating the NDVI at effective cover and for dry, bare soil at the experimental site to the K*cb* at effective cover and for dry soil evaporation, respectively. Similarly, [22] obtained (4) for two research sites in Colorado using alfalfa as reference evapotranspiration surface

$$K_{cb} = 1.181 \cdot NDVI - 0.026 \tag{4}$$

In order to minimize the presence of soil background, other vegetation indices (VI) have been used to compute K*cb* [23]. This provides a particularly useful tool for satellite images where soil brightness and colour can vary. One of these VIs used is Soil Adjusted Vegetation Index (SAVI) [24] which for the same conditions as (4) gives the relationship [5]:

$$K_{cb} = 1.69 \cdot SAVI - 0.16 \tag{5}$$

Thus, crop coefficients derived from spectral measurements (K*cs*) are independent of the time parameters, day of planting and effective cover, and represent a real-time crop coefficient. The use of spectral crop coefficients facilitates irrigation scheduling on a field-to-field basis over a large region if the fields can be observed spectrally, because planting and assumed effective cover dates are not required.

The spectral information would be sensitive to leaf loss due to hail, stress caused by disease and water deficit, cold or wet conditions that delay early growth, and warm temperatures and drought that speed senescence [16]. At field scale, further work was performed in order to improve scheduling irrigation events on corn compared to other traditional DoY based methods resulting in estimated crop water use reduced by 15% [25].

5. CROP COEFFICIENT AND NDVI RELATION FROM FIELD OBSERVATIONS

Intensive experimental campaigns were conducted within pilot zones. The research field has a permanent lysimeter station and it is water controlled following FAO 56 specifications [26]. Coinciding in time with spectral acquisitions, biomass ($kg.m^{-2}$), Leaf Area Index (LAI), and Green Fraction Cover (*fc*) were measured to describe the phenology of crops.

By the knowledge of crop stages, K*cb* values have been estimated taking into account the effect of varying relative humidity and wind velocity from standard conditions (RH = 40%, v = 2 $m.s^{-1}$) (Allen et al., 1998). Reflectance in red and near infrared to compute NDVI is obtained by integrating spectral reflectance in the range of ranges [0.63, 0.69 m] and [0.76, 0.90 m].

Evolution in time of NDVI and *fc* for maize is represented in Figure 5. NDVI reaches its maximum value, when crop reaches also full effective green cover in coincidence to maximum of K*cb*. As can be seen in Figure 3 for maize, ranges of maximum and minimum values for *fc* and NDVI coincide in time obtaining comparable curves.

Variation in behaviour of *fc* allows determining K*cbINI* (0.15), and K*cbMID* (1.15). To determine K*cbEND* it is necessary to estimate water content of plant. The resulting 54% on DoY = 277 suggest a value of K*cbEND* = 0.5. The average curve for K*cb* adapted for crop height, minimum relative humidity and wind velocity is represented in Figure 5 along with NDVI.

From linear regression we obtain the equations for the reflected-based crop coefficients for corn (Figure 6) in case of NDVI and SAVI:

$$K_{cbNDVI} = 1.37 \cdot NDVI - 0.017 \qquad \left(R^2 = 0.99\right) \tag{6}$$

$$K_{cbSAVI} = 1.76 \cdot SAVI - 9.10^{-2} \qquad \left(R^2 = 0.99\right) \tag{7}$$

To perform the comparison between (6) and (7), which are grass based reference evapotranspiration, and (4), which is alfalfa based reference evapotranspiration, we have realized the following steps:

1. Derive from (4) a new SAVI based equation, using the relationship NDVI = 1.2735•SAVI+0.02106, obtained from the definition of SAVI with a value for L = 0.5 (where L is the adjusting factor to account for *fc* in the SAVI definition equation), the same as used for (7);
2. Multiply alfalfa-based K*cb* by a factor 1.15 to convert them in grass based K*cb*, according the procedure described in [27]. Equations (8) and (9) show that the obtained results are very similar to (6) and (7).

$$K_{cb,grass} = 1.36 \cdot NDVI - 0.031 \tag{8}$$

$$K_{cb,grass} = 1.73 \cdot SAVI - 0.009 \tag{9}$$

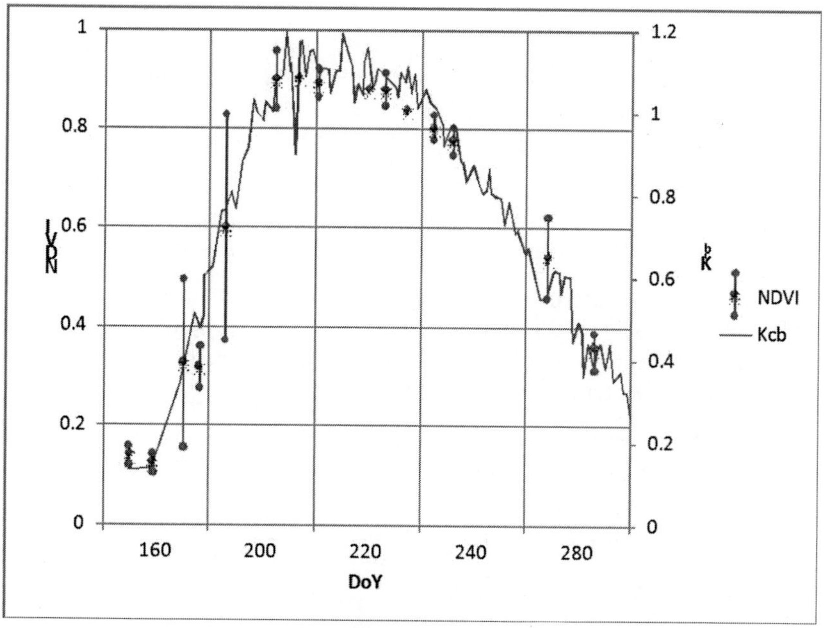

FIGURE 5. Field observations of Kcb and NDVI versus DoY for corn.

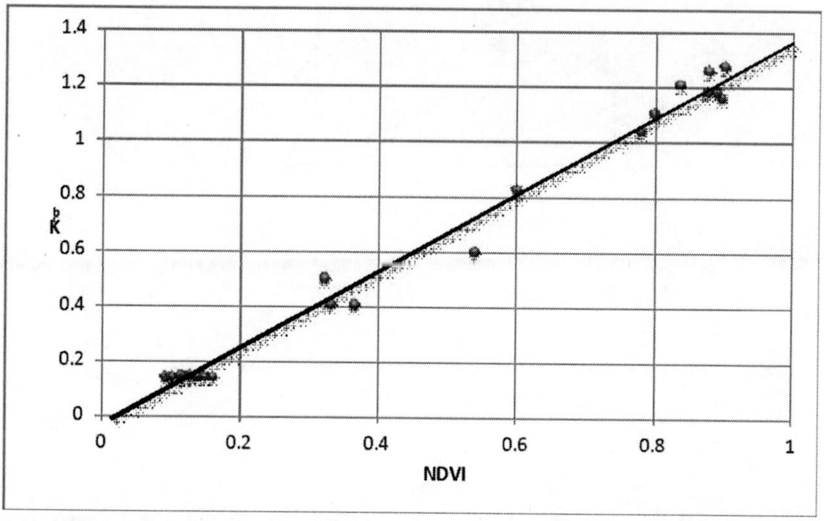

FIGURE 6. Linear regression between values of K*cb* and measured NDVI for maize.

In the case of wheat, the evolution in time is not as representative as in maize. We observe in Figure 3 that on DoY = 95, *fc* reaches a first local maximum before it continues growing in coincidence with emerging ears (with active photosynthesis), supposing a rapid increase in fc, and thus in K*cb* and spectral indices.

Applying (6) and (7) to data obtained for wheat, we see in Figure 7 that K*cb* obtained from NDVI reproduces the evolution in time of fc.

This relationship facilitates calculations of transpiration taking into account that only points over dry soil were considered, but without limiting crop transpiration. Evaporation of soil introduces an important contribution to K*c* during days after irrigation or rainfall.

This means that water soil balance must be taken into account to get the contribution of evaporation in K*c*. Over large areas, where variability of soil colour and brightness can influence the NDVI, SAVI and other Vis designed to normalize soil background effect should be used.

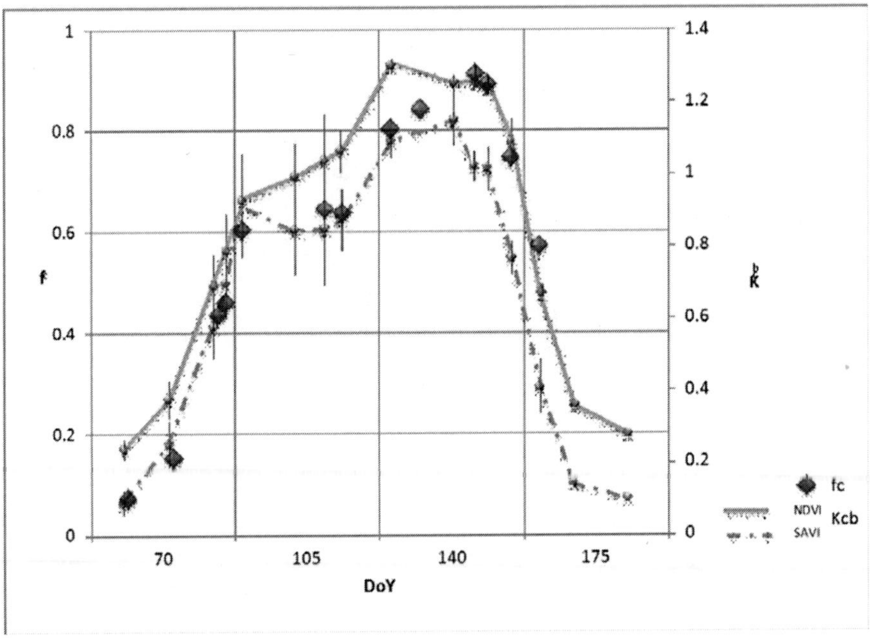

FIGURE 7. Evolution in time of observed fc for wheat. The K*cb* NDVI and K*cb* SAVI values for wheat have been obtained from the linear relationships in (6) and (7).

6. CROP COEFFICIENT FROM NDVI: OPERATIONAL POINT OF VIEW

Equations (6) and (7) provide the grass based basal crop coefficient from NDVI and SAVI data. These VIs are calculated for Landsat TM broadband from field radiometry data. Equation (4) provides alfalfa-based basal crop coefficient from NDVI.

Neale et al. [28] review the use of canopy reflectance observations to obtain crop coefficients over large areas. Similarities were found between the mean crop coefficient for small grain to the ratio of the perpendicular vegetation index (PVI) for wheat to PVI of wheat at full canopy cover. Heilman et al. [3] investigated the relationship between percent cover and reflectance-based perpendicular vegetation index (PVI) for alfalfa.

Neale et al. [22] related the crop canopy reflectance to basal crop coefficient for corn, developing an operational technique for estimating actual crop ET. The reflectance based crop coefficient (Kcr) was derived by nearly transforming the seasonal normalized difference vegetation index (NDVI) using the percent shading and leaf area measurements to establish the EFC and relate it to the basal crop coefficient by [6]. In several studies, NDVI has been directly used to predict Kc [5, 22, 23, 29, 30].

The operational procedure to estimate Kc from satellite imagery is based on the linear relationship between NDVI and basal crop coefficient described earlier. Landsat is the reference imagery to estimate NDVI (spectral broadband calibration). The attractiveness of Landsat is the high resolution (30 m in the visible and near infrared bands and 60 to 120 m in the thermal band) so that individual fields can be observed. The methodology that is described here has been checked against the preliminary results of all PLEIADES pilot areas for the following crops: Barley, wheat, maize, opium plant, sugar beet, alfalfa, pea, potato, onion and garlic. So we can establish the limits of applicability of this approach.

6.1. Dual Crop Coefficient Ndvi Approach

Wright [6] proposed a dual basal crop coefficient approach which splits the total crop coefficient into crop transpiration (Kcb) and soil evaporation (Ke) fractions, see Figure 8. The Kcb component represents the crop evaporative conditions from soil conditions whose surface is dry (direct evaporation from soil surface is minimum), and the crop growth is not limited by water, insect, climatological or physiological factors. The dual crop coefficient concept (see also equation (2)) expressed as

$$K_c = K_{cb} + K_e \tag{10}$$

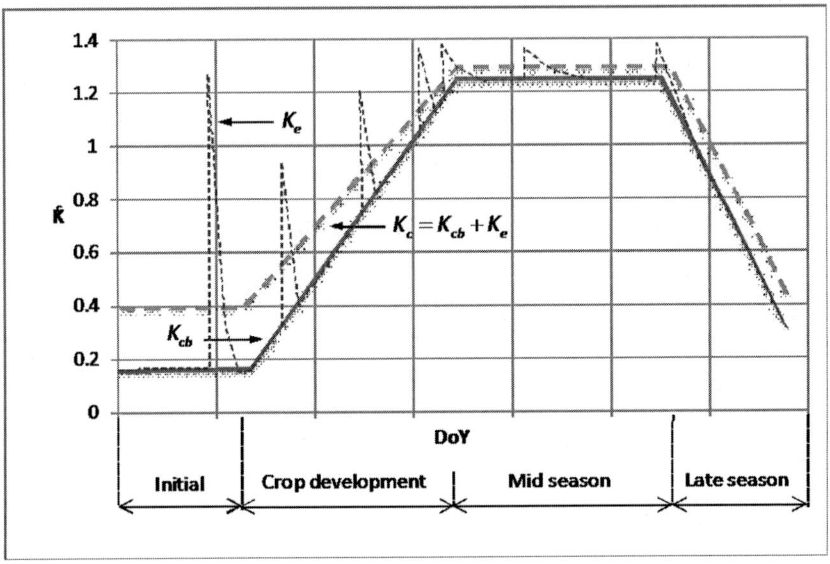

FIGURE 8. Crop coefficient curves showing the basal K*cb*, soil evaporation K*e* and the corresponding single K*cb* = K*cb* + K*e* curve [1].

We assume that there is a linear relationship between K*cb* and NDVI as stated earlier. This time, however, the linear relationships are adjusted to values of NDVI*MAX* and NDVI*MIN* from satellite imagery rather to those from field radiometry. Table 2 gives the corresponding values.

TABLE 2. Maximum and minimum values of NDVI and derived parameters observed in set of Landsat images.

	NDVI	K_{cb}	f_c	K_c
Minimum	0.16	0.15	0.0	0.4
Maximum	0.80	1.15	0.8	1.2

The resulting linear relationship is:

$$K_{cb} = 1.5625 \cdot NDVI - 0.1 \quad (11)$$

The soil evaporation part in (10), K*e*, is related with bare soil fraction, and is strongly dependent on wetting state of bare soil fraction, because the evaporative power of soil changes strongly if the soil is wetted or if the soil is dry. Irrigation system (gravity, sprinkler, drip, etc) and irrigation frequency, coupled with type and stage of crop, are the factors that determine the time of different bare soil wetting states. We propose a first approach to take into account these factors assuming *NDVI* as a good estimator of ground fractional cover, fc, (and so, of bare soil fraction, 1 - fc). The other factors are parameterized by means of a parameter β:

$$K_e = (1 - f_c) \cdot \beta \quad (12)$$

The parameter β is estimated empirically, from the values of K*cINI* or K*cMID* and can be modified on the basis of ancillary or local information. It is crop (and stage) dependent. Assuming a linear relationship between *NDVI* and fc for all crops, and considering again the *NDVI* maximum and minimum values from satellite imagery and the corresponding fc as given in Table 1, we obtain the relationship

$$f_c = 1.3514 \cdot NDVI - 0.2811 \quad (13)$$

6.2. SINGLE CROP COEFFICIENT NDVI APPROACH

A common β parameter value is 0.25, obtained considering an fc value of 0.8, K*c* equal to 1.2, and K*cb* equal to 1.15. Taking β as 0.25 and combining (10), (11), (12), and (13), we obtain a direct relationship K*c* - *NDVI*

$$K_c = 1.2246 \cdot NDVI + 0.2203 \quad (14)$$

We also obtain a relationship K*c* - *NDVI* directly from Table 1 in the same way as above, by considering a linear relationship between the maximum *NDVI* and the maximum K*c* (at effective full cover) and the minimum (bare soil) *NDVI* and bare soil Kc, respectively. The resulting relationship is:

$$K_c = 1.25 \cdot NDVI + 0.2 \tag{15}$$

Equations (14) and (15) are very similar. By its simplicity we assume (15) as the operational formula to derive Kc from *NDVI*.

Table 3 shows the comparison between Kc values obtained from NDVI by means of (15), and the values for Kc*INI*, Kc*MID* from [1] for the crops studied in the field during the pilot campaign. We observe good agreement for crops with higher effective ground cover, although (15) seems to overestimate Kc*INI* slightly for spring crops.

TABLE 3. Comparison between the averaged Kc values obtained from NDVI by means of (15), and the values for Kc*INI*, Kc*MID* from FAO56 [1] for the main crops in the pilot zone.

Crop	NDVI min	NDVI max	K_{cINI} (15)	K_{cINI} FAO56	K_{cMID} (15)	K_{cMID} FAO56
Alfalfa	0.16	0.80	0.40	0.40	1.20	1.20
Barley	0.16	0.80	0.40	0.30	1.20	1.15
Garlic	0.16	0.44	0.40	0.70(0.40)*	0.75	1.00
Maize	0.16	0.78	0.40	0.30(0.40)*	1.17	1.20
Onion	0.16	0.53	0.40	0.70(0.50)*	0.86	1.00
Opium poppy	0.16	0.80	0.40	No reference	1.20	No reference
Pea	0.16	0.77	0.40	0.50(0.40)*	1.16	1.15
Potato	0.16	0.78	0.40	0.50(0.45)*	1.17	1.15
Sugar beet	0.16	0.78	0.40	0.35(0.45)*	1.17	1.20
Wheat	0.16	0.80	0.40	0.30	1.20	1.15

Significant deviations between NDVI based and FAO56 based Kc*MID* are found for crops like garlic and onion, which exhibit low ground cover in the stage of maximum development, in contrast with the rest of crops studied. The higher bare soil proportion of those crops can introduce and reinforce effects on Kc related with irrigation frequency, irrigation system utilized, environmental aspects and others. Further research is under way to study this behaviour.

It should be also noted that (14) and (15) are applicable for the initial crop development and mid-season phases only. The application for the late season phase, when the crop is maturing, requires a slight correction because ground cover (green and dry) remains nearly constant in that phase. Assuming a constant value of 0.8 for ground cover (Table 1) and combining (11) and (12), we obtain:

$$K_c = 1.5625 \cdot NDVI - 0.05 \tag{16}$$

Summarizing, the operational equations will be (15) for the initial, development and mid season and (16) for the late season.

7. CAIA (PORTUGAL)

The Caia irrigation scheme is located in the Caia watershed in east-central Portugal close to the Spanish border. The Caia river is a tributary to the transnational Guadiana river basin. It is a relatively flat area at a mean elevation of 200 m above sea level, with mean annual precipitation of 537 mm. The irrigation infrastructure was established in 1967, with a central dam of 203 hm^3 capacity and a metered canal network serving 7,237 ha. Except for a concentrated small-scale plot area of 600 ha, the fields are generally large, on average 35 ha. The main crops are corn, wheat, tomato, and sugarbeet, which are all grown for industrial commercialization.

The water resources from the dam are mainly used for irrigation (over 90%) and population supply (less than 10%). Pressure on water quantity is increasing and water quality is already under pressure, as ecological standards are very close to the regulatory limits.

The Associação de Beneficiários do Caia (ABCaia) is the local water user association (Irrigation District Board). It has the mandate for water management in the irrigation scheme and represents the local farmers in the River Basin Council. At the beginning of PLEIADES, there was an incipient traditional IAS (provided by the Centro Operativo de Técnicas de Rega, COTR) with a newly installed agrometeorological station and a small GIS facility at the Irrigation District Board. The goal was to build up an innovative irrigation advisory capacity, in order to cope with the rapidly increasing pressure on water quantity and quality.

7.1. Local Crop Expert Database and Field Protocols

The first pilot campaign was a fruitful training and learning phase that consolidated the local team and established local field sampling protocols. It also laid the foundations for the local GIS-based expert database on crop phenology. Figure 9 shows the crop coefficient curves for the major crops. An important emphasis in the Caia pilot zone is on tomato, which has phonological cycles that can vary enormously between plots.

During the following pilot campaign, extensive field data were collected to extend the expert database for all major crops of the area. They were also used for validation and local calibration of EO-derived products.

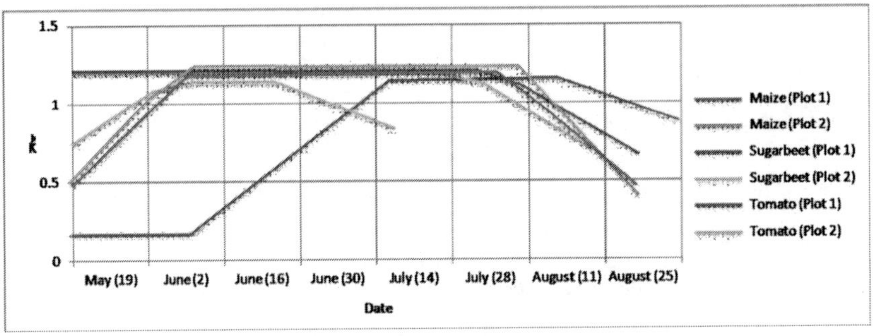

FIGURE 9. Crop coefficient curves for major crops in Caia area, from field campaign.

The local team developed and implemented a work strategy having as objective to be close to satellite overview conditions. In that sense nadir pictures of crop canopy have been taken as close as possible to overpass satellite period, to determine Green fraction cover, and the crop coefficient has been calculated using farming information concerning amount of used irrigation water, agrometeorological data and field phenological stage to correct and calibrate K*cb* curve.

For this campaign a spatial extrapolation of the crop coefficients was performed, based on field work land use maps. Figure 10 shows the temporal and spatial evolution of crop coefficient in the pilot zone, for the three studied crops.

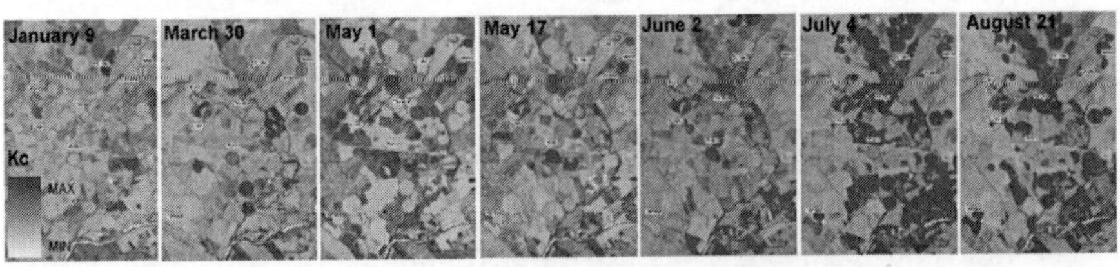

FIGURE 10. Spatial and temporal evolution of crop coefficient in Caia pilot Zone.

7.2. Calibration and Validation of Fao and Eo Derived Crop Coefficients

The crop coefficient has been calculated in three different ways: the first (line in Figure 11) is based on the concepts of the FAO-56 (Allen et al., 1998) methodology for *Kcb* calculation, with its general tabulated values; the second is Kc *NDVI* calculated with *EO* PLEIADES methodology (circles and squares, from Landsat 7 and 5, respectively); third way is *Kcb* from field observations, adjusted with additional field data (triangles). Figure 11 shows maize monitored on "Melinho" and "Botafogo" test fields. The length of the "Initial stage" of phenological development is based on information concerning seeding dates, but FAO-56 standard methodology stipulates that crop green development occurs earlier than has been observed in the field. The EO-derived data are rather close to the field observations. The slight scatter indicates that the crop stage have several variations, depending on soil and crop water stress parameters.

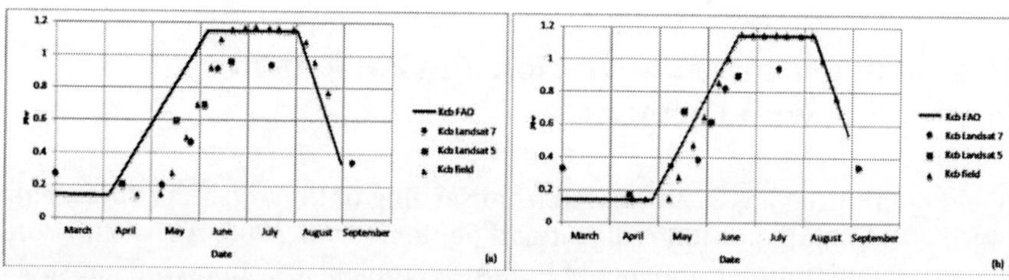

FIGURE 11. Maize field K*c* evolution during campaign versus EO derived and FAO-56: Melinho plot area (a) and Botafogo plot area (b).

Tomato is planted in a way that full canopy is not reached in any growing stage. This means that soil evaporative fraction is always present, resulting in a better water management at plot level. Due to the fact that it is a multi-stage plant, i.e. one plant can be on multiple phenological stages at the same period, irrigation must be a compromise for the average stage of the field. Based on the experience of field technicians, the different *Kcb* curves (Figure 12) have been calculated with data obtained at plot level, FAO-56 methodology and *EO* derived data.

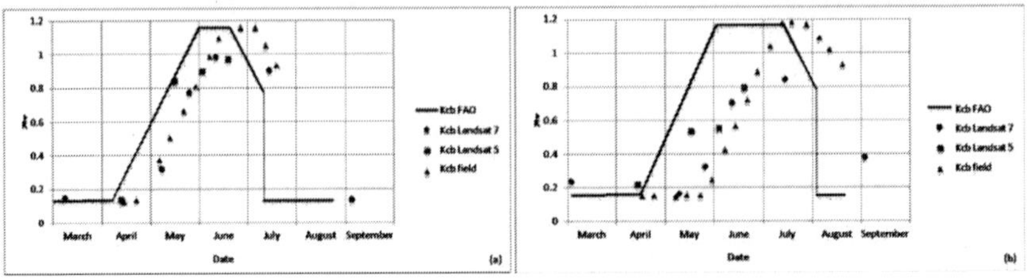

FIGURE 12. Tomato field K*c* evolution during campaign versus EO derived and FAO-56: D. Joana plot area (a) and Sta. Isabel plot area (b).

The results demonstrate again a huge difference between standard FAO-56 values (line) and the K*c*values obtained from field survey (triangles). As with maize, the EO derived data present closer values to what has been observed on the plots. Given the huge deviation of actual crop coefficients from FAO-56 values the local expert database on crop phenology plays a crucial role in the quality of an*IAS* under implementation.

7.3. Real-Time Demonstration and Participatory Evaluation with Farmers

Field monitoring has been carried from seeding until harvesting dates, only during the months of July, August and September of 2005. A real-time on-place demonstration campaign of PLEIADES took place, surveying hot spots in each of the 6 pilot fields and delivering *IAS* information to farmers.

A "Farmer Report" was supplied to each individual land owner covered by the ground truth operation (Figure 13). In Figure 13 is clearly identified the difference in corn maturation in July, using drop-to-drop irrigation (a) and pivot irrigation (b).

Several meetings were organized to promote and explain the content of the information to the farmers and to get their feedback. The farmers found that the EO derived information, along with field survey data, can be very useful for them and that such an irrigate advisory service can help them with irrigation and farming strategies at their farm holding level.

Although, one of the monitored fields had a change for sugar beet with a winter variety, with impacts on data correlation, on the other two the same season of maize and tomato has been used.

The weekly procedure has been maintained in order to collect phenological stages data to define K*cb* field curves, for evaluation of EO-derived data. For the demonstration campaign the Portuguese Meteorological Institute provided Kc *NDVI* data, obtained from Landsat 5, in almost real time, so that they could be compared with field data. The local team made their spatial integration and data extrapolation based on ABCaia GIS tools and related geographical information data available for this pilot zone.

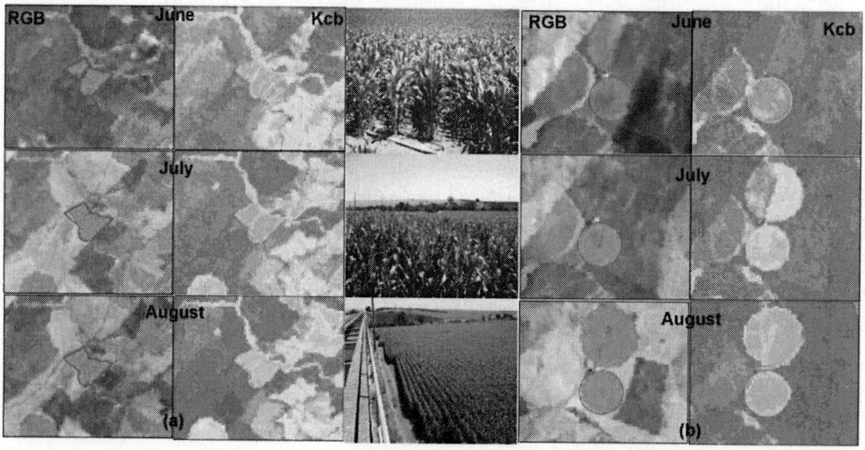

FIGURE 13. Farmer report from D. Isabel (a) and Melo (b) fields.

Knowing, through data from the meteorological station at Caia (Figure 14), the relationship between potential evapotranspiration (ET*p*) and evapotranspiration measured from *EO* (*ETEO*), it is possible to establish the relationship expressed in (2).

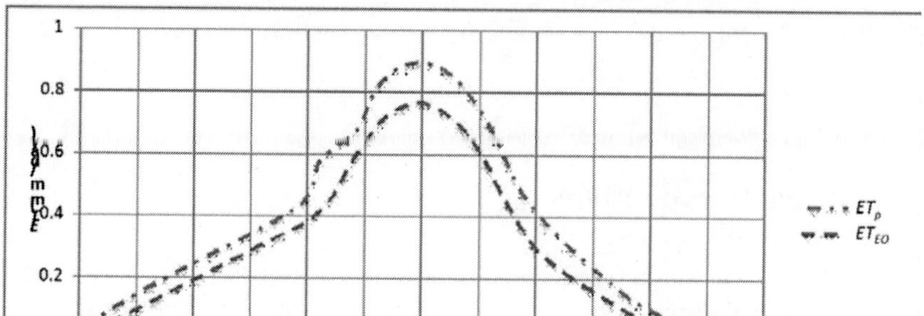

FIGURE 14. Evapotranspiration in Caia area.

Hence, having a value of, for example, $ET0 = 7.2$ for the 2nd of June we can make the correspondence for the all area (Figure 15). This kind of data is very important to farmers because there is a well known relation where precipitation plus irrigation is equal to evapotranspiration plus terrain drainage.

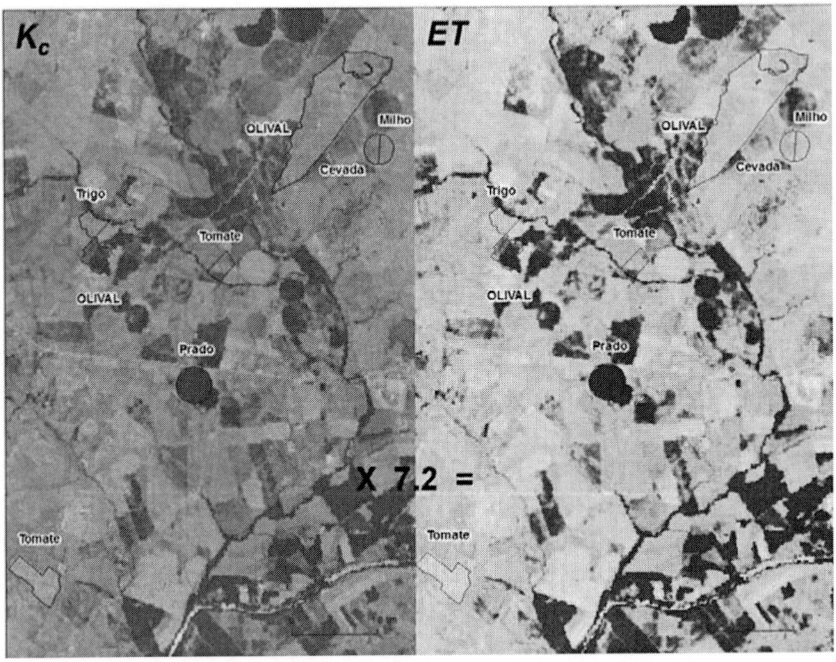

FIGURE 15. Crop coefficient and Evapotranspiration in Caia area at 2 of June.

7.4. Future Perspectives

For the intensive and extensive campaign the local ground team has been organised and defined protocols and set up a strategy to collect data as close as possible to satellite overpass conditions. During the project period a local effort has been made to implement irrigation advisory services on the pilot zone area, resulting in ongoing use of agrometeorological station data provided by the Centro Operativo de Técnicas de Rega (COTR) and the Farmer Association (ABCaia) supplying information on irrigation parameters. Survey data have been available due to the cooperation of the Agronomy School of Elvas University (ESA) and also made available to them jointly with with ABCaia aiming to set up strategies for development of local technical skills to support irrigation development in accordance with environmental requirements.

The foundations have been laid for the PLEIADES prototype to start being operative in the Caia pilot zone. An efficient and dedicated local operating team has been consolidated, with collaborating entities. A local *GIS* based expert database on crop phenology and EO methodology calibration has been developed and the necessary ICT infrastructure has been established (Figure 15). The user participation is incipient but promising. A funding model still needs to be found for further sustainable implementation. Increasing pressure on water quantity and quality may provide an important motivation.

The Server is based on leading edge online GIS Technology. The system architecture is web based and composed by a modular group of components, which makes maintenance of the system easy. Those components are based on the XML language.

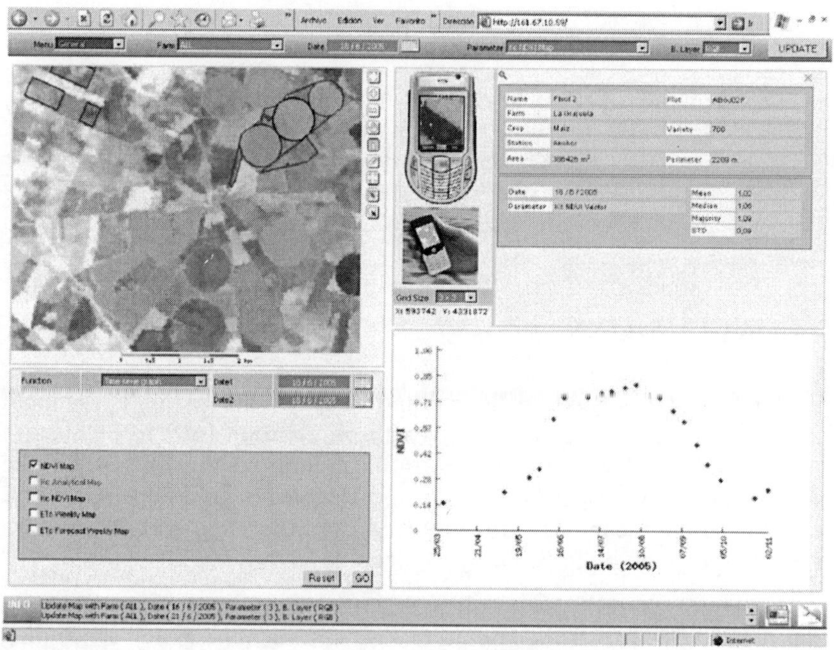

FIGURE 16. PLEIADES WebGIS and mobile applications example.

The web services have been programmed also to let the server be distributed in different PCs to share the load of the system. The server has been programmed for a Windows system, using open source libraries and toolkits like FOP, Xerces C++, GDAL, GD, MapServer, SWFF, etc. GDAL has been used for obtaining the satellite images information, getting the pixel value and creating charts with GD showing the time evolution.

This toolkit, as well as the queries form, helps the user in taking decisions about water, crops, etc. MapServer lets the system offer a visual support for the user, supporting a wide variety of image formats, as well as vector format and databases. Xercess is the parser for the XML messages that implements the DOM interface providing quite an easy interface for the programmer for accessing the XML information.

The system receives information from images and stations, pre-processing some of them to provide formatted data. It also offers a Quality Control (QC) interface for the IAS manger to control the information for the farmer, so that the final information distributed to the user ("published") includes only correct and accurate information. The system is prepared for assimilating the satellite image as far as possible, so that the system gives a real time toolkit for analysing the satellite information with the spatial information.

The mobile phone client has been programmed in Flash. A simpler interface has been designed because of the limitations of the mobile technologies. This client lets the farmers access the information of the satellite wherever they are without needing a computer or an internet connection and with an easy interface that lets them zoom into the image or ask for parameters information.

8. CONCLUSIONS

PLEIADES was designed to assess and demonstrate in an operational perspective how the integration of Earth observation (*EO*) techniques in routine Irrigation Advisory Services (IAS) can improve the efficiency in the use of water for irrigation. The use of leading-edge Information and Communication Technology (*ICT*) tools in the generation and distribution of information makes the *EO* easily available to *IAS* and the farmers.

The PLEIADES WebGIS (www.pleiades.es) is the central outcome of the project. Its key feature is the operational generation of irrigation scheduling information products from a virtual constellation of *EO*satellites and their delivery to farmers in near-real-time using leading-edge on-line analysis and visualization tools. It is supported by a methodology package to derive crop coefficients and further advanced parameters from *EO* satellite images in an operational processing chain.

PLEIADES basic products were generated and transmitted to a sample of farmers normally within 2 days from overpass, thus completely matching the weekly operational irrigation scheduling cycles.

Participatory evaluation with selected farmers shows that the farmers feedback is very positive, both on the information quality as on the added value of the spatial information (within-plot heterogeneity and between-plot variations). The reliability and accuracy of the information has been con-firmed by the comparison of different approaches to derive crop coefficients from EO and validation with field data in all pilot zones.

The major improvement achieved by the use of *EO* in the generation of basic IAS information products like crop coefficients is twofold. Firstly, the spatial coverage is enhanced significantly, both extending to larger areas and providing within-field heterogeneity information. Secondly, the spatially resolved *EO* data can easily be combined with cadastral information in a geographical in-formation system (*GIS*), which allows for personalization of the irrigation scheduling recommendation.

Conventional IAS provides average irrigation recommendations per crop type, while the new space-assisted IAS is able to provide specific recommendations for each individual plot, based on the actual state of that plot.

The fast image delivery and quality controlled operational processing make the EO-based crop coefficient maps available at the same speed and quality as ground-based data (point samples), while significantly extending the spatial coverage and reducing service cost. The uptake of users at *IAS* and farmer level is encouraging.

Advanced products have made a significant step towards operationality while maintaining satisfactory levels of accuracy. First exploitation steps including full operational implementation are indicators of the success of the prototype and the project.

The space segment is the most vulnerable part of the entire operational system. After the sensor failure of Landsat 7, the backbone of the actual system is Landsat 5, due to its excellent operationality and low cost (22 years old, with no replacement in sight). Urgent actions are required to ensure the capability to obtain adequate EO images at the adequate coverage frequency and low cost.

REFERENCES

1. R. G. Allen, L. S. Pereira, D. y. Raes, M. Smith, 1998 Remote Sensing Based Crop Coefficients for Water Management in Agriculture UN-FAO, Rome, Italy.
2. J. O. Bailey, 1990 Remote Sensing Based Crop Coefficients for Water Management in Agriculture. Remote Sensing Reviews, 4 2 349377 .
3. J. L. Heilman, W. E. Heilman, D. G. Moore, 1982 Evaluating the Crop Coefficient Using Spectral Reflectance. Agronomy Journal, 74 967971 .
4. R. G. Allen, A. Morse, M. Tasumi, 2003 Application of SEBAL for western US water rights regulation and planning. University of Idaho, Kimberly, Idaho, USA. Proceedings of ICID-CIID International Workshop: Use of remote sensing of crop evapotranspiration for large regions: September 17, Montpellier, FRANCE.
5. B. J. Choudhury, N. U. Ahmed, S. B. Idso, R. J. Reginato, C. S. T. Daughtry, 1994 Remote Sensing Based Crop Coefficients for Water Management in Agriculture. Remote Sensing of Environment, 50 117 .
6. J. L. Wright, 1982 New Evapotranspiration Crop Coefficients, Journal of Irrigation and Drainage, 108 1 5774 .
7. A. Calera, C. Martínez, J. Meliá, 2001 Remote Sensing Based Crop Coefficients for Water Management in Agricultureley. International Journal of Remote Sensing, 22 17 3357 EOF3362 EOF .
8. T. N. Carlson, D. Ripley, 1997 Remote Sensing Based Crop Coefficients for Water Management in Agriculture. Remote Sensing of Environment, 62 241252 .
9. P. J. Sellers, R. E. Dickinson, A. Randall, A. K. Betts, F. G. Hall, J. A. Berry, G. J. Collatz, A. S. Denning, H. A. Mooney, C. A. Nobre, N. Sato, C. B. Field, A. Henderson-Sellers, 1997 Modelling the Eschanges of Energy, Water, and Carbon between Continents and the Atmosphere. Science, 275 52-509.
10. J. L. Monteith, 1977 Climate and Efficiency of Cropproduction in Britain. Phil. Trans. Res. Soc., London, SeriesB. 281, 277329 .
11. G. Asrar, M. Fuchs, E. T. Kanemasu, J. L. Hatfield, 1984 Estimating absorbed photosynthetic radiation and leaf area index from spectral reflectance in wheat. Agronomy Journal 76 300306 .

12. S. T. Gower, C. J. Kucharik, J. M. Norman, 1999 Remote Sensing Based Crop Coefficients for Water Management in Agriculture. Remote Sensing of Environment, 70 2951 .
13. A. Calera, J. González-Piqueras, J. Meliá, 2004 Remote Sensing Based Crop Coefficients for Water Management in Agriculture International Journal of Remote Sensing,, 25 1 97 EOF109 EOF .
14. J. L. Monteith, M. H. Unsworth, 1990 Remote Sensing Based Crop Coefficients for Water Management in Agriculture., E.Arnold Ed., London, 290
15. M. A. Gilabert, S. Gandía, J. Meliá, 1996 Analyses of Spectral-Byophysical Relationships for a Corn Canopy. Remote Sensing of Environment, 55 1120 .
16. W. C. Bausch, C. M. U. Neale, 1987 Remote Sensing Based Crop Coefficients for Water Management in Agriculture Transactions of the ASAE,30 3 703709 .
17. N. J. Rosenberg, B. L. Blad, S. B. Verma, 1983 Microclimate ; The biological environment 2nd ed. New York ; John Wiley & sons, 1315 ,170.
18. A. Calera, A. Jochum, A. Cuesta, 2003 Earth observation techniques to improve Irrigation Advisory Services. Proc. Envirowater, Albacete, 46 September.
19. A. M. Jochum, A. Calera, A. Cuesta, 2003 Irrigation management from space: Towards user-friendly products. Proc. ICID Workshop "Use of Remote Sensing of Crop Evapotranspiration fro Large Areas", 17 September, Montpellier. (submitted Int'l J. Irrigation & Drainage)
20. M. S. Moran, Y. Inoue, E. M. Barnes, 1997 Remote Sensing Based Crop Coefficients for Water Management in Agriculture. Remote Sensing of Environment, 61 319346 .
21. W. C. Bausch, C. M. U. Neale, 1987 Remote Sensing Based Crop Coefficients for Water Management in Agriculture Transactions of the ASAE, 30 3 703709 .
22. C. M. U. Neale, W. C. Bausch, D. F. Heermann, 1989 Remote Sensing Based Crop Coefficients for Water Management in Agriculture. Trans. ASAE 32 6 18911899 .
23. W. C. Bausch, 1993 Remote Sensing Based Crop Coefficients for Water Management in Agriculture. Remote Sensing of Environment, 46 213222 . 46:213-222.

24. A. R. Huete, 1988 Remote Sensing Based Crop Coefficients for Water Management in Agriculture. Remote Sensing of Environment, 25 295309 .
25. W. C. Bausch, 1994 Remote Sensing Based Crop Coefficients for Water Management in Agriculture. Agricultural Water Management, 27 5568 .
26. J. Gonzalez-Piqueras, A. Calera, M. A. Gilabert, 2003 Remote Sensing Based Crop Coefficients for Water Management in Agriculturengs of the SPIE, Barcelona, September, 811 th.
27. ASCE Hydrology Handbook 2 Ed. (1996). Chapter 4: Evaporation and Transpiration. ASCE Manuals and Reports on Engineering Practice 28
28. C. M. U. Neale, H. Jayanthi, J. L. Wright, 2003 Crop and irrigation water management using high resolution airborne renote sensing. Utah State University, Logan, Utah, USA. Proceedings of ICID-CIID International Workshop: Use of remote sensing of crop evapotranspiration for large regions: September 17, Montpellier, FRANCE.
29. W. C. Bausch, C. M. U. Neale, 1989 Remote Sensing Based Crop Coefficients for Water Management in Agriculture. Transactions of the ASAE, 32 6 19011908 .
30. W. C. Bausch, 1995 Remote sensingof crop coefficients for improving the irrigation scheduling of corn. Agricultural Water Management. 27 5568 .

4

CHAPTER

STUDENT PERCEPTIONS OF PROBLEM-BASED LEARNING: A CASE STUDY OF UNDERGRADUATE APPLIED AGROMETEOROLOGY

Linda De Wet[1] and Sue Walker[2]

[1]Department of Soil, Crop and Climate Sciences, University of the Free State, Agriculture Bldg, 2nd Floor, Office 1.230, Bloemfontein 9301, South Africa
[2]Crops for the Future Research Centre (CFFRC) Level 2 Block B, The University of Nottingham Malaysia Campus Jalan Broga, 43500 Semenyih, Selangor Darul Ehsan, Malaysia

ABSTRACT

Many students do not seem to transfer their learning during formal education into applications in the real world. The objective of this ongoing study was to investigate the opinion of third-year students concerning their program through problem-based learning and to improve the module where necessary. Students attending theory classes had to apply their newly gained knowledge coupled with real-life weather data to solve a problem during practicums. Students attending practicums were given the same questionnaire thrice; thus, the answers were based on different sets of exercises. Responses by attendees for the three questionnaires were 73%, 100%, and 61%, respectively. Students preferred problem-based practicums (78%, 54%, and 72%, resp.) to other non-problem-based practicums. Most students thought that their knowledge had improved and it had prepared them better for the workplace (85%, 77%, and 92%, resp.). Generally students preferred working in groups (74%, 62%, and 56%, resp.), in contrast to those preferring to work individually. Students benefited from problem-based learning in that they thought they had improved their knowledge, skills, and critical thinking abilities and felt that they had learnt things that they could carry into their future lives out in the world at large and the workplace.

1. INTRODUCTION

Traditional methods of teaching and learning have been used for centuries in almost all countries of the world and are therefore a universal phenomenon [1]. According to Kelly and Finlayson [2], teaching the traditional way is a teacher-centred approach where students follow a strict set of instructions, with perhaps little understanding of the purpose thereof, and in the case of practicums, many students are not able to produce satisfactory interpretation of the observations or results. According to Kendal-Wright and Kusuya [3], this "one-way communication" method of teaching has often been challenged and is not popular because students are passive learners. Using the problem-based approach, however, the emphasis is on students learningsomething they can apply outside the classroom and not on the teacher merely giving information [4] that the student does not even remember.

Problem-based learning was originally developed medical staff at McMaster University in Canada in the 1970s, and since then many others in different fields have based their own curriculum development on this model [5]. According to Dochy et al. [6], problem-based learning was developed in order to help students in the development of professional skills such as problem-solving, analysis, synthesis, and evaluation, as well as the ability to solve problems in the real world.

According to Kendal-Wright and Kasuya [3], problem-based learning starts out with a problem scenario, which requires that a number of learning topics be identified and that the student revisits the material acquired until it has been mastered. Problem-solving skills are therefore developed during the problem-based learning process [3], but one of the goals of problem-based learning is to allow several and not only one single solution [7]. Students actively participate, while identifying learning gaps, researching new knowledge, and devising plans and their own success. Students therefore learn independently, as well as from others in their group [7].

Problem-based learning has various advantages and disadvantages. Strohfeldt and Grant [8] list some of the advantages as follows.(1)The student learns in a more natural way.(2)Critical thinking is encouraged.(3)A good relationship between student and facilitator is encouraged.(4)The student spends on average five to eight hours in preparation for each session, whereas no preparation is done for traditional lectures [8].

Some of the disadvantages are, however, as follows [8].(1)Staff costs are high as problem-based learning is a staff-intensive process.(2)High staff workload, especially for initial preparation of new classes.(3)Facilitator expertise may vary, perhaps resulting in lack of high quality facilitation for all problem-based classes.

Strohfeldt and Grant [8] mention that students often complain of high workloads, but deep learning is best embedded in this way. According to Rust [9] and Biggs [10] students commonly approach learning in one of two ways, that is, a surface or a deep approach. With the surface approach students memorize isolated facts [9]. Students following the deep approach feel the need to engage on a deeper level, trying to understand the bigger picture but in the process also learning more detailed facts. Greasley and Ashworth [11] and Richardson [12] agree that a deep approach to learning includes reading for meaning and understanding. Richardson [12] however, states that there are three approaches to learning: surface approach to memorize for assessment; strategic approach based on obtaining highest marks; and deep approach based on understanding.

According to Strohfeldt and Grant [8], their way of presenting the problem-based learning course addresses one of the greatest problems, that is, high workload for staff, but each institution is free to adapt the model according to their specific needs. Students who normally study alone appreciate the help from group members. Students also learn not to cram, and at-risk students are the ones who mostly benefit [3]. Other authors, such as Zingone et al. [13], also promote mixed active learning (both individual and groups), besides team-based learning in groups.

In the Faculty of Natural and Agricultural Sciences [14] of the University of the Free State (UFS), Bloemfontein, South Africa, there are two streams of undergraduate programs; one of the streams falls under natural sciences and the other under agriculture. One of the syllabi under agriculture is agrometeorology wherein the module LWR314, called "The Influence of Climate on Agricultural Practices," is presented. UFS is also the only university in South Africa that offers a degree in agrometeorology to undergraduate students (Lomas et al. [15]; Lomas et al. [16]). In 2003 the LWR314 was redeveloped as a problem-based learning module by the subject head of agrometeorology to help students develop their professional skills and to expose them to problem-solving skills, which they could use not only in the classroom but in the real world as well. In order to improve the module it was decided to investigate student perceptions of the problem-based module, and so a simple questionnaire was developed in which students could voice their opinions thrice during the semester. Various other studies [6, 17, 18] used a 4- or 5-point Likert scale as, measuring instrument handed out at the end of the semester, but the questionnaire used in this study made use of a descriptive research analysis approach [19] as the research was aimed at looking at the progress of students throughout the semester. The objective of this study was to investigate the perceptions of students with regard to problem-based learning in the undergraduate applied agrometeorology class and to use the information on their views to further improve the module.

2. METHODOLOGY

The questionnaire, a one-page sheet consisting of eight questions, was voluntarily and anonymously answered three times during the semester, observing suggestions by Babbie [20]. Mouton [21] also states that sensitive or threatening questions could result in nonresponse or refusal to participate. During 2011, third-year students attending the LWR314 practicum class were each given the questionnaire at three of the thirteen practicum sessions. The first questionnaire was given on 18 March, the second on 6 May, and the last one on 13 May. The same questionnaire was used although the students' answers would be based on different sets of practicum exercises. These were, for the sake of simplicity named P1 (first survey date), P2 (second survey date), and P3 (third survey date). Responses to the first questionnaire (P1) were based on the foregoing seven practicum topics: discomfort indices, short-term weather forecasting, weather and remote sensing, evapotranspiration methods, calculation of heat and cold units, pests and diseases, and frost occurrence along a transect. Responses to the second questionnaire (P2) were based on the following three practicum topics: microclimate manipulation, irrigation scheduling, and fire indices. Responses to the third and final questionnaire (P3) were based on the last three practicum topics: drought mitigation, seasonal forecasting, and wind chill indices. There were three main lecturers presenting the practicums, with two guest lecturers who each presented one practicum exercise, both in the second group.

Figure 1 shows the questionnaire given to students in 2011. Analysis of the answers varied; for example, questions 1 and 2 were straightforward "Yes" (Y) or "No" (N) answers, but the rest of the questions involved some explanation or comment. The straightforward questions, including the instances where students did not answer (indicated by "None"), were summarised descriptively using frequencies and percentages. Similar explanations/comments were grouped together [19] and analysed within and between the questionnaires. Comments which were not repeated were not included, but reference to some of them was made in the discussion of results. Two of the questions, that is, questions 3 and 6, were compared to each other to determine whether a specific method of working was more prevalent under those who preferred problem-based learning. Question 3 asked which method of working (individual, group, or both) students preferred and question 6 asked if students preferred problem-based learning to non-problem-based learning. An ANOVA analysis was done to test whether there were significant differences between the survey dates (P1, P2, and P3) for percentages of all yes and no questions (questions 1, 2, 5, 6, 7 and 8). The null hypothesis was assumed; that is, there were no significant differences between the three survey dates.

LWR314 Question sheet 3 13 May 2011

LWR314

The influence of climate on agricultural practices

Practicals (drought, seasonal forecasting, windchill)

Please take a few minutes to answer the following questions honestly. We need your input so that we can improve on the classes and practicals.

Please cross the appropriate box and write answers on the dotted lines.

..............................oooOOOooo..............................

(1) Could you easily access the excel data on blackboard? [Y] [N]

(2) Did you read the instructions for the practical before you attempted to do it?

[Y] [N]

(3) Did you work in a group or individually or both?..............................

Which do you prefer?.............................Why?..............................

....

(4) What is the most important thing/lesson you have learnt from this practical?

....

....

(5) Do you feel that you benefited from the practical? | Y | | N |

How?..

....

(6) Do you prefer this type of (problem-based) practical to practicals which are not problem based? | Y | | N |

Why?..

....

(7) Do you think the practical has improved your knowledge and better prepared you for the work place? | Y | | N |

Why?..

....

(8) Do you have recommendations for improving the practical? give your reasons ....

....

Thank you for your time and effort!

Figure 1: Survey questionnaire given to third-year students at the University of the Free State (2011).

3. RESULTS AND DISCUSSION

Forty-one students were registered for the LWR314 module during the first semester of 2011. Of these, 27 of the 37 students who attended (according to signed roll call) completed the P1 questionnaire. The response was thus 73% of attendees and 66% of the total number of registered students. Twenty-six students completed the P2 questionnaire although only 22 students attended the practicum (four students completed the questionnaire outside of the practicum time as they were unable to attend this particular practicum session). The response was 100% for attendees and 63% for the total number of registered students. Questionnaire P3 was not handed out during the practicum session, as was the case with P1 and P2, but was completed by students who came to collect their semester test results a few days later. The P3 questionnaire response was 61% of the total number of registered students.

3.1. Answers to Yes/No Questions

The analyses of yes/no questions for P1, P2, and P3 are given in Table 1. The majority of answers from question 1 asking if students could easily access the Excel data on Blackboard were in the affirmative, with high percentages of 70% (P1), 85% (P2), and 96% (P3) answering yes. Therefore, this shows that as the semester progressed the students became more familiar with accessing data from Blackboard and it could also be interpreted that the students had acquired the necessary skills for using a spreadsheet for this type of problem-based exercise. In contrast the "no" decreased so that all except one student was familiar with this process by the end of the semester.

Table 1: Analysis of questionnaires for "Yes/No" answers for P1, P2, and P3 (P1 = 1st survey date; P2 = 2nd survey date; and P3 = 3rd survey date).

Question no.	Question	P1 (n = 27)			P2 (n = 26)			P3 (n = 25)		
		Yes n (%)	No n (%)	None n (%)	Yes n (%)	No n (%)	None n (%)	Yes n (%)	No n (%)	None n (%)
1	Could you easily access the *Excel* data on Blackboard?	19 (70)	7 (26)	1 (4)	22 (85)	4 (15)	—	24 (96)	1 (4)	—
2	Did you read the instructions for the practical before you attempted to do it?	27 (100)	—	—	26 (100)	—	—	23 (92)	1 (4)	1 (4)
5	Do you feel that you benefited from the practical?	22 (81)	4 (15)	1 (4)	22 (85)	3 (12)	1 (4)	21 (84)	4 (16)	—
6	Do you prefer this type of (problem-based) practical to practicals which are not problem based?	21 (78)	5 (19)	1 (4)	14 (54)	10 (38)	2 (8)	18 (72)	6 (24)	1 (4)
7	Do you think the practical has improved your knowledge and better prepared you for the workplace?	23 (85)	4 (15)	—	20 (77)	5 (19)	1 (4)	23 (92)	2 (8)	—
8	Do you have recommendations for improving the practical?	11 (41)	8 (30)	8 (30)	13 (50)	6 (23)	7 (27)	14 (56)	5 (20)	6 (24)

Results for question 2, which asked if the instructions were read before the practicum was attempted, also indicated 100% yes answers for P1 and P2, but in P3 one student said no and another did not answer the question (Table 1). Possible reasons for students not answering these first two questions could have originated from the fact that they read the questionnaire hastily instead of slowly and thoroughly. This phenomenon is often observed in students who feel they have too much to do with too little time, and this is in accordance with findings by Strohfeldt and Grant [8].

Question 5, which asked if students felt that they had benefited from the practicums, showed yes results to be in the majority, with 81% in P1, 85% in P2, and 84% in P3 (Table 1). The mean of students who felt they did not benefit was 14%, and probably they were disinterested in this particular subject matter. Only one student in each of P1 and P2 did not answer the question. Perhaps this question was only seen by the student to mean "benefit" in this module, whereas actually one would hope that at least some of the students had learnt some transferable skills that would benefit them in other modules or areas of their lives. According to Forsythe [4], transferable skills include time management, team work, independent learning, decision taking, problem solving, and communicating ideas, in other words, things that would be applicable to other areas in life and not only in the sphere of academics.

Question 6 asked if students preferred these practicum types to non-problem-based practicums, and results once again indicated that the majority of students answered in the affirmative (78% (P1), 54% (P2) and 72% (P3)) (Table 1). It is probably important for lecturers in other courses to take note of this result, that students feel that they learnt more from "problem-based" classes and to try to include such studies into their classes also. An average of 30% said they did not prefer the problem-based practicum, and 1 or 2 students preferred to give no answer to the question at all. The low result of 54% for the middle survey date at P2 may have something to do with (a) the fact that classes were presented by guest lecturers during the time that the P2 questionnaire was given; (b) the midterm tests; and (c) the lower number of students who actually attended that specific practicum class.

Question 7 asked if students thought that their knowledge had improved and better prepared them for the workplace (Table 1). Most of the students felt that this was the case (85% (P1), 77% (P2) and 92% (P3)), while 15%, 19%, and 8%, respectively, felt that it was not the case. Only 1 student from P2 did not answer this question. Of course these answers are only subjective as these students have not yet entered the workplace as such, because they have not yet finished their full-time studies or graduated.

Lastly, question 8 asked if there were any recommendations for improving the practicums (Table 1). Students who said no were 30% (P1), 23% (P2), and 20% (P3), respectively, while 30%, 27%, and 24% did not answer the question. It was surprising that this is the only question where the "no" and "none" responses have similar numbers of students, almost as if some 30% of the students do not want to participate in the formulation of the course or they feel that is not their responsibility, and so they show disinterest by not even responding to this question. Despite this, the highest response to question 8 is still with the students answering a "yes" indicated by 41%,

50%, and 56%, respectively (Table 1). It is also interesting that this proportion of the class increased as the semester progressed, and this shows that although the majority agreed that the method of learning and teaching is good, they still have helpful suggestions to improve the class further.

3.2. Answers to Questions 3 and 4

Question 3 firstly asked if students worked in groups or individually or both and secondly, which they preferred. At all three survey dates (P1, P2, and P3), students preferred working in groups (74%, 62%, and 56%, respectively—not shown), although the percentages from P1 to P3 show a decrease in this preference. Those preferring to work individually increased from 15% in P1 to 19% in P2 and 28% at the last survey date (not shown). Perhaps this shows that due to the pressure of time towards the end of the semester, the students rather completed their tasks alone or that they had become familiar with the methods used in the problem-based learning and preferred to work alone. Students who preferred both also showed an increase up to 15% in P2 and 16% in P3 from a mere 7% in P1. It seems that the students formed working groups during the first half of the semester and then some of them left these groups as the semester progressed. These students obviously learnt independently as well as from others in their groups. This phenomenon is in agreement with other literature sources such as that by Benedict [7], who reported in an advanced therapeutic study that the medical students had learnt independently as well as from each other.

Table 2 represents a summary of the comments from all three survey dates, and the results overall indicate that students were generally very positive about this problem-based module. The results in Table 2 were very difficult to analyse because the students express themselves differently resulting in many different answers, but an effort was made to group similar comments together; for example, in question 3, students gave their comments about why they preferred to work on their own or individually, in a group or both.

At all three survey dates (P1, P2, and P3) the term "less confusion" was coined for comments such as the following.

I. I like to be certain.

II. More can be achieved (students most probably mean they can concentrate better when they are not confused).

III. I understand the work better.

IV. I get less confused.

V. Mistakes can be more easily corrected.

VI. I like to solve problems.

VII. I know what I know.

Table 2: Summary of main comments, according to question number on questionnaire given to LWR314 students at all three survey dates (P1, P2, and P3) (Number of students who made comments on P1, P2, and P3, respectively, given in brackets under comments).

Question no.	Question	Comments
3	Do you prefer to work in a group or individually or both? Why?	Individually—less confusion (3, 3, 3) and less conflict (0, 3, 3). Group—understanding improved (12, 8, 8). Both—interaction aspect (3, 3, 4).
4	What is the most important thing/lesson you have learnt from this practical?	How Excel works (6, 2, 1). Knowledge of the weather (9, 4, 0). Knowledge of drought (0, 0, 4). Critical thinking (0, 3, 3).
5	Do you feel that you benefited from the practical? How?	Yes—Knowlege accumulation (4, 7, 6); learnt applicability. (skills/practical application possible) (3, 7, 4); understanding improved (0, 3, 3). No—No understanding (2, 0, 2); not applicable (1, 0, 0); programs difficult (0, 1, 0); too many practicals (0, 1, 0).
6	Do you prefer this type of (problem-based) practical to practicals which are not problem based? Why?	Yes—Critcal thinking (5, 3, 2); learnt applicability/skills (3, 1, 1); reenforces understanding (1, 2, 2). No—It is difficult (1, 3, 4); no reason (2, 4, 0); confusing (0, 2, 0).
7	Do you think the practical has improved your knowledge and better prepared you for the workplace? Why?	Yes—Critical thinking (2, 1, 3); prepares for workplace (5, 0, 6); knowledge accumulation (2, 5, 2); learnt applicability/skills (1, 2, 2). No—No understanding (2, 2, 2).
8	Do you have recommendations for improving the practical? Give your reasons.	Yes—More tutors (2, 0, 1); more patient lecturers (1, 0, 1); preparation before practical (2, 2, 2); memorandums (2, 2, 0); more examples (0, 4, 8). No—No recommendations (8, 6, 6). None—No answer (8, 7, 6).

Another reason for preferring to work alone was that there was less conflict. This comment was given in P2 and P3 but did not occur in P1, probably because this may have been the first time students had worked in groups in a science subject and some found that it had not really worked for them (because of possible conflict). Students who preferred working in groups (Table 2) said that their understanding had much improved with number of students stating this type of answer equal to 12, 8, and 8 in P1, P2, and P3, respectively. This is in accordance with statements made by Rust [9] who stated that one should remember that research evidence indicates that students need to actively participate in classes in some way (such as in a group format) to be able to understand what is being taught. Kelly and Finlayson [2] stated that a combination of group work, discussions, and hands-on activities allowed students to develop their understanding through deep learning. Students who preferred working both individually and in groups numbered 3 for P1 and P2 and 4 for P3, the reason being that they liked the interaction with the other students, as then they could receive input and/or help from the group as well as give their own input, but still do the practicum on their own. It is assumed that they may have worked in groups during the initial stages of an exercise then alone during the final write-up for submission. Alternatively, perhaps they initially worked alone but struggled too much, so they reverted to group work to solve the problem and do the analysis. To be able to distinguish between these alternatives one would have to either interview the students or provide more detailed questions at a later stage, so it is not possible to decipher now.

Question 4 asked what the most important thing/lesson learnt was, and students mostly acknowledged that they had acquired a knowledge of the weather with 9, 4, and 0 students giving this comment in P1, P2, and P3, respectively. During each of the questionnaire opportunities, however, students appreciated that they did learn the spreadsheet package Excel (6, 2, 1, resp.), and in P2 and P3, three students in each said they had learnt how to think critically. Presumably, these skills will be transferable and will stand them in good stead in future. It is very encouraging to know that they could even acknowledge that they learnt it in this agrometeorology class. A study conducted by Kumar and Kogut [19] stated that the goal of learning is to encourage students to be reflective critical thinkers. Masek and Yamin [22] also state that various studies show positive results relating problem-based learning and critical thinking, although their own study did not show significant differences between problem-based and conventional learning. In this study students were required to give their feedback on problem-based learning, and some of the comments were that the students felt they had become more independent as they learnt to think, rather than being "spoon-fed."

3.3. Answers to Questions 5 and 6

Question 5 asked whether and in what way the students had benefited from the exercises in the practicums. Table 2 shows that the students who answered in the affirmative mostly said that they had accumulated knowledge and learnt applicability or skills (practical application possible) in the sphere of weather and climate. Some also said their understanding of weather- and climate-related matters had improved. Students who answered in the negative said that they did not understand the practicums (P1 and P3) or that the practicums were too difficult or that there were too many (P2). One student in P1 also said that the practicums were not applicable to him/her, so one wonders why that student was taking this module. Students could have accumulated knowledge by other means/methods of teaching such as through rote learning; however, deep learning and understanding and being able to apply the things learnt to other situations is seldom learnt by "rote" or memorization or by repetition. According to Kumar and Kogut [19] student feedback suggested that students preferred problem-based learning to rote learning as the latter meant having to sit through long exam sessions answering questions from memory rather than making their own inferences. Hilgard et al. [23] also found that students who learn with understanding are able to transfer their knowledge to tasks requiring problem-solving with greater success than those who learn only by rote.

Question 6 asked students about their preferences in learning methods comparing problem based practicum exercises to those which were not problem based. Table 1 clearly shows that the majority preferred the problem-based practicum exercises, and it would be reasonable to assume that they are not just trying to appease the lecturer but are giving an honest answer. Reasons given were that their critical thinking was stimulated (Table 2); applicability or skills were learnt, and understanding was positively reenforced. These answers were recorded at all three survey dates through the semester. Students who answered in the negative said that the work was difficult (1, 3, and 4 in P1, P2, and P3, resp.), while others said they had no reason (2 and 4 in P1 and P2, resp.) or that it was confusing for them (2 in P2). So if it was confusing, this provides an opportunity for the lecturing staff to work on improving the introduction and explanation of the specific exercises and how to approach them. If the problem originates with the student, then maybe it is because the student does not remember/recall the information from introductory classes on agrometeorology which they took the previous year.

3.4. Answers to Questions 7 and 8

Question 7 asked whether the students thought the practicums had improved their knowledge and better prepared them for the workplace and why. Students answering in the affirmative said they were confident that they were prepared for the workplace (5 in P1, 0 in P2, and 6 in P3), and one would hope that this was due at least in part to the learning method;

others, namely, 2, 5, and 2 (P1, P2, and P3, resp.), felt they had accumulated knowledge, which could probably have been achieved with other learning methods also. Some other reasons were also given, such as that their critical thinking had improved and that they had learnt how to apply what they were learning to real-life problems as well as new skills. Students answering this question in the negative said they did not understand the practicums (2 at each survey date). The possibility always exists that some students did not understand the practicums or that they missed the introductory lecture class or did not have the correct background to understand some of the technical terms. During this semester the lecturers were (and are always) willing to help them, but often there is no indication that there are students who do not follow. One can then only assume that the students do not want to admit their ignorance (particularly in front of other students), either because they are too shy or because they did not concentrate during the explanation of the practicum.

Question 8 asked if there were any recommendations for improving the practicum exercises. The majority of the students gave recommendations for improvements, but differences between the remainder who said no and those who did not comment was quite small; that is, in P1 eight students said no and 8 students did not answer. In P2 six students said no and 7 did not answer. In P3 six students said no and 6 did not answer (Tables 1 and 2). So it seems as though either these students did not want to apply their mind to this side of the class and how it could be improved or that they were pushed for time when completing the survey and could not bother to think about alternatives or other applications or things that they would see as problems that could be developed into learning modules. Some of the recommendations from the majority of students were as follows (Table 2).

I. More tutors (2 in P1 and 1 in P3).

II. More patient lecturers (1 in P1 and 1 in P3).

III. Preparation before practicums (2, 2, and 2, resp.).

IV. Memorandums to be given and discussed (2 each in P1 and P2).

V. More examples to be given (4 and 8 in P2 and P3).

Some of these recommendations need to be addressed before the class is presented next year. For instance, more tutors in the computer lab is really something that needs attention as has been stated, the number of students in this class had rapidly increased over the past few years, and perhaps lecturers had not given this enough thought and did not consider that the tutor-to-student ratio had now become quite low. This needs to be brought to the attention of the head of department, and it needs to be emphasised that in order to maintain the course in this "problem-based" learning style more tutors are necessary or the students become frustrated, as they must wait too long before receiving assistance during the 3-hour practicum class time. The recommendation about impatient lecturers may also be due to work overload and pressure during this time with over 35 students and only 2 people to help answer questions when each student is working on a computer as an individual. It is a little unclear what is meant by "more preparation before the practicum"—is this by the student or by the lecturer? Some students recommended that lecturers discuss and explain the practicum during the week before the practicum session (but actually the lecturer usually does this—so maybe it is due to poor class attendance) to help save them time during the practicum session itself. The last two comments are mainly due to the time pressure during each week for each topic with only 3 × 1 h classes and 1 × 3 h practicum period—so no extra time is available for other examples or to go over the answer sheets/memorandum, only after the assignment has been graded and returned to the students.

There were several other recommendations made, both positive and negative, which were not repeated at all survey dates, for example, in P1: "Keep up the good work—we enjoy it"; or in contrast in P2: "Just cancel the practical—no good at all," and in P3: "Everything on hardcopy," and so forth. These recommendations were also taken note of as each student is entitled to his or her own opinion, but the latter comment is totally impractical (pages and pages of data) and would not be in line with keeping up to date, technologically speaking.

3.5. Comparison of Questions 3 and 6

According to the results of the comparison between questions 3 and 6 for the yes answers (Figure 2), the majority of students (59%) in P1 chose group work as their preferred method of working in a problem-based learning environment. This was compared to 36% of students in P3 and to 31% of students in P2. In P3, 20% of students said they preferred to work individually, but 16% of students from P3 and 15% from P2 said they preferred both. These results, therefore, indicate that approximately one-third to two-thirds of the students who said they preferred problem-based practicums also said they preferred group work. This is very encouraging as one of the requirements for problem-based learning is that students work in groups, and students can really benefit if they are prepared to communicate with others. White (2002), as quoted by Kelly and Finlayson [2], stated that students would understand better what they learned and "remember longer by working cooperatively in groups,"

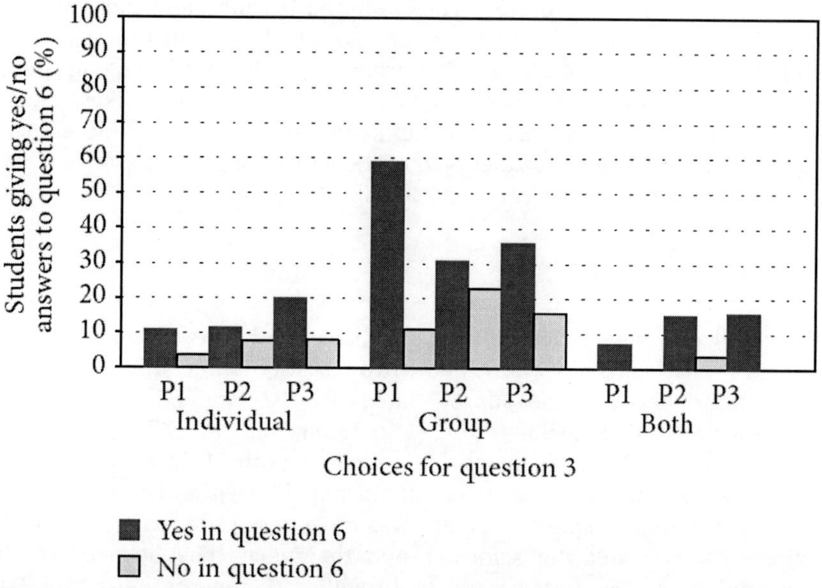

Figure 2: Comparison of results for questions 3 (do you prefer working in a group or individually or both?) and 6 (do you prefer problem-based practical or not?) for the questionnaire completed by third-year students at survey dates P1, P2, and P3.

These group working skills would also be transferable to other situations that the students find themselves in outside the classroom. A minority of students (<7%) did not respond (not shown) to the question at all three survey dates (P1, P2, and P3).

According to Figure 2 most of the students who said no to question 6 (did not prefer problem-based learning) showed 23% for group work in P2. In other words, these students enjoyed working in groups but did not prefer a problem-based learning environment. This was compared to 16% students in P3 and 11% of students in P1. All other results were ≤8%, that is, very few students who did not prefer problem-based learning but preferred to work individually, or both, or none. This implies that overall most of the students who did not prefer problem-based learning also preferred group work. Students therefore seemed to prefer the interaction with other students, even though they shied away from the problem-based type of teaching. The goal of the lecturers should be to try and narrow the gap between students who prefer problem-based learning and those who do not; that is, students should be taught how to think—in this way more students will come to prefer solving problems, as opposed to being spoon-fed. Students who do not prefer problem-based learning most likely are not very successful in this area, and this needs to be taken into account by lecturers by encouraging students to participate in problem-based learning, so that learning outcomes can be optimal [19].

3.6. Statistical Analysis

Results from the ANOVA analysis performed on the answers in Table 1 (questions 1, 2, 5, 6, 7, and 8) to test for significant differences in percentages between the survey dates (P1, P2, and P3) showed that all values > 0.05. The values were: (a) 0.78 for "yes," (b) 0.20 for "no," and (c) 1.00 for "none", which means that the null hypothesis is accepted, and so this indicates that there were no significant differences between survey dates. This seems to show that there was no or very little change in student opinions throughout the semester and that what they believed in the beginning of the semester continues through to the end of the course. This points to relative consistency in the percentages between the survey dates; for example, question 5 asked whether students felt they had benefited from the practicums, resulting in 81%, 85%, and 84% "yes" answers, which is stable through the semester. There would be a serious problem had the percentages been very different (and hence statistically significant) because then one would have to further investigate the cause of the inconsistency and whether it was due to student or lecturer error or oversight. Even though the outcome for the ANOVA analysis was statistically insignificant, the results are by no means unimportant [24].

4. CONCLUSIONS

The objective of the study to investigate how students felt about learning through problem-based learning has been achieved. The study showed that the overwhelming majority of students in the 2011 LWR314 class were in favour of problem-based learning, which is based on making applications of theory into practical situations. The main reason students gave for preferring to work in groups was that their understanding had improved. Students who preferred to work on their own felt there was less confusion and conflict, and those who preferred both enjoyed the interaction that groups provided as well as being able to formulate their own individual ideas. Students benefited from problem-based learning in that they improved their knowledge, skills, and critical thinking abilities and felt that they had learnt something that could benefit them throughout their future lives out in the world and the workplace. The lecturing staff at agrometeorology (UFS) are determined to give each student the best assistance and guidance possible and to improve the problem-based module LWR314 where necessary, taking into consideration the comments and results collected in this study. Case studies such as this one prove the importance of allowing students to develop their critical thinking skills. These skills will remain with them even after they have left the university environment.

ACKNOWLEDGMENTS

The study was presented at the CALMet IX conference in Pretoria from 3 to 8 October 2011. The authors would like to acknowledge Liesl van der Westhuizen (University of the Free State) for assistance with the survey as well as editing of the paper. S. Walker is the research theme leader for crop physiology, agroclimatology, and agroecological potential and programme director of CropBase, CFFRC.

REFERENCES

1. M. Motah, "Learning, types of learning, traditional learning styles and the impact of e-learning on the performance of secondary school students: the perceptions of teachers," in Proceedings of the Computer Science and IT Education Conference, University of Technology, Mauritius, India, November 2007.
2. O. C. Kelly and O. E. Finlayson, "Providing solutions through problem-based learning for the undergraduate 1st year chemistry laboratory," Chemistry Education Research and Practice, vol. 8, no. 3, pp. 347–361, 2007.
3. C. Kendal-Wright and R. Kasuya, "Team based learning: a potential addition to the JABSOM curriculum," Hawaii Medical Journal, vol. 69, no. 10, pp. 247–248, 2010.
4. F. Forsythe, "Problem-based Learning," in The Handbook For Economics Lecturers, P. Davies, Ed., University of

Staffordshire, http://www.economicsnetwork.ac.uk/handbook/printable/pbl_v5.pdf.

5. M. Karakas, "Graduating reflective science teachers through problem based learning instruction,"Bulgarian Journal of Science and Education Policy, vol. 2, no. 1, pp. 59–71, 2008.
6. F. Dochy, M. Segers, P. V. D. Bossche, and K. Struyven, "Students' perceptions of a problem-based learning environment," Learning Environments Research, vol. 8, no. 1, pp. 41–66, 2005.
7. N. Benedict, "Teachers' topics. Virtual patients and problem-based learning in advanced therapeutics,"The American Journal of Pharmaceutical Education, vol. 74, no. 8, article 143, pp. 1–5, 2010.
8. K. Strohfeldt and D. T. Grant, "A model for self-directed problem-based learning for renal therapeutics,"The American Journal of Pharmaceutical Education, vol. 74, no. 9, article 173, pp. 1–7, 2010.
9. C. Rust, "The impact of assessment on student learning," Active Learning in Higher Education, vol. 3, no. 2, pp. 145–158, 2002.
10. J. Biggs, Teaching for Quality Learning at University, Open University Press, Buckingham, UK, 2nd edition, 2003.
11. K. Greasley and P. Ashworth, "The phenomenology of approach to studying: the university student's studies within the lifeworld," British Educational Research Journal, vol. 33, no. 6, pp. 819–843, 2007.
12. J. T. E. Richardson, "The role of response biases in the relationship between students'perceptions of their courses and their approaches to studying in higher education," British Educational Research Journal, vol. 38, no. 3, pp. 399–418, 2012.
13. M. M. Zingone, A. S. Franks, A. B. Guirguis, C. M. George, A. Howard-Thompson, and R. E. Heidel, "Comparing team-based and mixed active-learning methods in an ambulatory care elective course," The American Journal of Pharmaceutical Education, vol. 74, no. 9, article 160, pp. 1–7, 2010.
14. Faculty of Natural and Agricultural Sciences, Yearbook 2011. Part 4: Agricultural Sciences: Undergraduate Programmes, University of the Free State, Bloemfontein, South Africa, 2011,http://natagri.ufs.ac.za/templates/yearbook.aspx.
15. J. Lomas, J. Collins, S. Gachara, and S. Mukhopadhyay, "Education and training in Agrometeorology,"CAgM Report 78, WMO/TD No.

990, Geneva Switzerland, 2000,http://www.wmo.int/pages/prog/wcp/agm/publications/cagm_reports.php.

16. J. Lomas, J. R. Milford, and E. Mukhala, "Education and training in agricultural meteorology: current status and future needs," Agricultural and Forest Meteorology, vol. 103, no. 1-2, pp. 197–208, 2000.
17. K. Gregson, L. M. Romito, and L. P. Garetto, "Students' attitudes toward integrating problem-based learning into a D.D.S. Pharmacology curriculum," Journal of Dental Education, vol. 74, no. 5, pp. 489–498, 2010.
18. S. J. Wright, "Student perceptions of an upper-level, undergraduate human anatomy Laboratory course without cadavers," Anatomy and Science Education, vol. 5, pp. 146–157, 2012.
19. M. Kumar and G. Kogut, "Students'perceptions of problem-based learning," Teacher Development, vol. 10, no. 1, pp. 105–116, 2006.
20. E. R. Babbie, The Practice of Social Research, Wadsworth Publishing Company, Belmont, Mass, USA, 7th edition, 1995.
21. J. Mouton, How To Succeed in Your Master's and docToral Studies, Van Schaik Publishers, Pretoria, South Africa, 7th edition, 2004.
22. A. Masek and S. Yamin, "The impact of instructional methods on critical thinking: a comparison of problem-based learning and conventional approach in Engineering education," ISRN Education, vol. 2012, Article ID 759241, 6 pages, 2012.
23. E. R. Hilgard, R. P. Irvine, and J. E. Whipple, "Rote memorization, understanding, and transfer: an extension of Katona's card-trick experiments," Journal of Experimental Psychology, vol. 46, no. 4, pp. 288–292, 1953.
24. D. H. Robinson, L. O'Ryan, J. R. Levin, and D. Halbur-Ramseyer, "Does statistical language constitute a "significant" roadblock to readers' interpretations of research results?" Journal of Educational Psychology, vol. 93, no. 3, pp. 646–654, 2001.

5

CHAPTER

ESTIMATION OF GROUNDWATER RECHARGES IN ODEDA LOCAL GOVERNMENT AREA, OGUN STATE, NIGERIA USING EMPIRICAL FORMULAE

Oluseyi O. Adeleke [1,*], Victor Makinde [1], Ayobami O. Eruola [2], Oluwaseun F. Dada [2], Akintayo O. Ojo [1] and Taiwo J. Aluko [1]

[1] *Department of Physics, Federal University of Agriculture, Abeokuta 110001, Nigeria*

[2] *Department of Water Resources Management and Agrometeorology, Federal University of Agriculture, Abeokuta 110124, Nigeria*

ABSTRACT

Estimation and forecast of groundwater recharge and capacities of aquifers are essential issues in water resource investigation. In the current research, groundwater recharge and the recharge coefficient were determined through a case study using empirical methods applicable to the tropical zones. The related climatological data between January 1983 and December 2014 were collected from Ogun-Oshun River Basin Development Authority (OORBDA), Ogun State, Nigeria. Using empirical formulae. The results showed that groundwater recharge was 194.7 mm per year, evapotranspiration was 1296.2 mm per year, and the recharge coefficient was 20.2% for the study area. The result showed that about 11% of rainfall infiltrated the aquifer, 73% was lost to evapotranspiration, and 36% ended up as run-off. Correlation between climatic parameters and groundwater recharge showed the highest correlation between recharge and rainfall. Temperature, humidity, solar radiation and evapotranspiraton were obtained at the 0.01 significance level; the results of linear regressions proved that precipitation has a significant effect (with $R^2 = 0.983$) on estimated recharge.

Keywords: aquifer; evapotranspiration; groundwater recharge; recharge coefficient

1. INTRODUCTION

In arid and semi-arid regions, the search for water which are under increasing stress from the growing human population, poses a great challenge due to its scarcity [1]. Groundwater, as a dynamic system, is located beneath the Earth surface and moves under the control of many factors, which are influenced by forces that are dependent on hydrogeology, hydrology, and climatology [2]. Recharge, as one of the factors controlling the situation and fluctuation of groundwater, is an important parameter that needs to be assessed more fully. Recharge occurring in small and large scales, spatially and temporally, is influenced by several factors, such as meteorology, soil characteristics, geology, surface cover, slope, and depth of the groundwater level [3,4,5]. Groundwater recharge estimation from precipitation is an integral part of hydrology and hydrogeology [6].

Although, precipitation is the most important source of groundwater recharge [7] the accuracy of currently attainable techniques for measuring recharge are not completely acceptable.

Comparative analysis of empirical formulae to estimate recharge which is a prerequisite for groundwater resource management was carried out [8]. Estimation of groundwater recharge was derived from rainfall data obtained from Ogun-Oshun River Basin Development Authority (OORBDA) using three empirical formulae, which include the empirical methods of [7,9,10]. The result revealed that the three formulae used gave comparable results, hence their conclusion that any of the formulae can be used for groundwater estimation. Additionally, in order to validate the formulae, the groundwater recharge and capacity of aquifers in Malaysia was estimated using empirical formulae [9] and obtained a recharge of 326.39 mm was obtained for the area [11].

In measuring groundwater recharge, indirect methods are employed since there are no universally accepted standard methods. However, methods for determining the groundwater recharges have been classified into three groups namely: Physical Model method, obtained when the recharge is calculated from the base flow; Chemical model method, which is used when the measurements of water soluble substances are considered; and Numerical model methods, such as HELP, RORA, PART, WEAP, WTF, PUISE, HYSEP and so on, which can be used in the estimation of groundwater recharges for various climates such as in arid, semiarid, and tropical regions albeit each with its own advantages and disadvantages [12].

2. THE STUDY AREA

Odeda Local Government Area is located on the outskirts of Abeokuta, an ancient city in the southwestern part of Nigeria, extending from longitude 3.3899E to 3.47950E and latitude 7.1745N to 7.2300N. Abeokuta is located in Ogun State, Southwestern Nigeria, as shown in Figure 1.

The study area is characterized by a tropical climate with distinct wet and dry seasons. The wet season is associated with the relative prevalence of the Moist Maritime Southerly Monsoon from the Atlantic Ocean, while the dry season is predicated by the Continental North Easterly harmattan winds from the Sahara desert. The area is located within a region characterized by a bimodal rainfall pattern (commencing in March; plentiful in July and September, with a short dry spell in August). The long dry season extends from November to March. Annual rainfall ranges between 1400 and 1500 mm in Abeokuta and environs. The region is characterized by relatively high temperature with a mean annual air temperature being about 30 °C. The greatest variation in temperature is experienced in July (25.7 °C) and in February (30.2 °C). Humidity is lowest (37%–54%) at the peak of dry season in February and highest at the peak of the rainy season between June and September (78%–85%).

The study area is underlain by basement complex, and the basement complex rocks of Nigeria are well-represented in the selected study area of Ogun State. These rocks are of Precambrian to early Palaeozoic age and they extend from the northeastern part of Ogun State running southwest and dipping towards the coast [13].

The basement complex metamorphic rocks are characterized by various folds, structures of various degrees of complexity, faults and foliation. These structural features have a predominant North-South or North-North-East-South-South-West (NNE-SSW) orientation which is particularly strong within the low grade metamorphic. The common metamorphic rocks encountered are gneiss, schist, quartzite and amphiboles. The study area is characterized by various rock types ranging from, granite granitic gneiss and pegmatite. The individual rocks have various hydro-geologic characteristics and belongs to the stable plate which was not subjected to intense tectonics in the past. Therefore, the underground faulting system is minimal and this has contributed to the problem of underground water occurrence in this area.

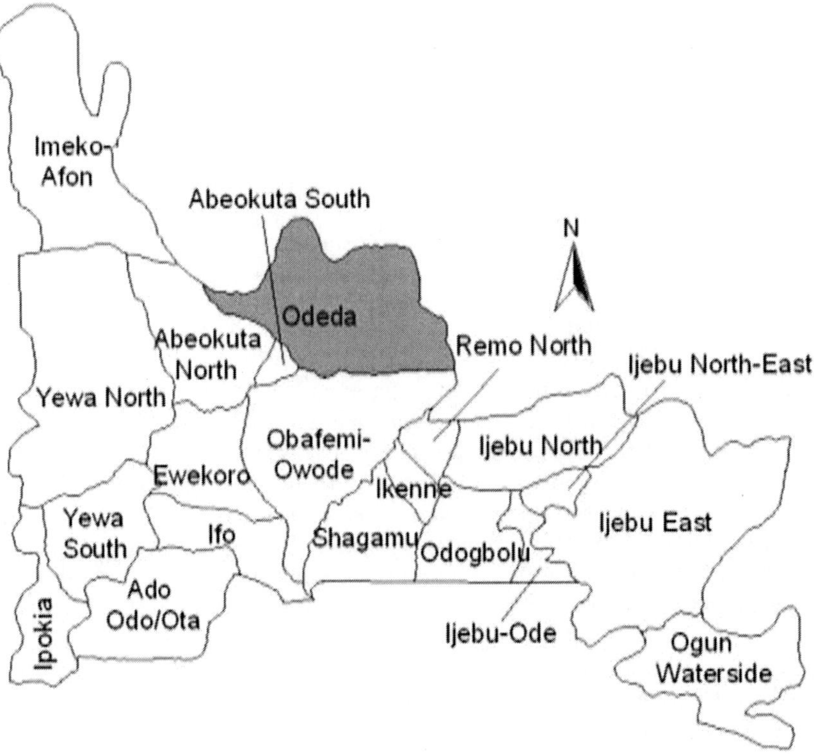

Figure 1. Location Map showing the study area.

3. METHODOLOGY

Owing to the non-availability of meteorological stations in the study area, Ogun Oshun River Basin Development Authority (OORBDA) meteorological station was used; parameters collected from OORBDA weather station were; minimum and maximum temperature, sunshine hour, relative humidity wind speed, and rainfall data.

Thirty-two (32) years of weather parameters were collected from OORBDA weather station. These were used to calculate the recharge, evapotranspiration, recharge coefficient and run-off using empirical formulae.

The potential evapotranspiration (PE) was determined using CROPWAT version 8.0 based on FAO Penman-Moteithe formula.

3.1. Evapotranspiration

Evapotranspiration is regarded as to the sum of evaporation and transpiration. It is a climatic index integrating the effect of air temperature, humidity, wind speed, and solar radiation. CROPWAT software version 8.0 was used to estimate evapotranspiration. The software is based on the FAO Penman-Monteith formula [14].

3.2. Estimation of Groundwater Recharge

Estimation of groundwater recharge of the study area was conducted using a modified version of [9] for tropical regions based on water level fluctuation and rainfall depth [14,15]. The equation is given as:

$$R = 1.35(P - 14)^{0.5} \quad (1)$$

where R is the net recharge due to precipitation in inches, and P is the precipitation in inches. These can be converted to millimeters (mm).

3.3. Recharge Coefficients

The value of the recharge coefficient is defined as the ratio of recharge to effective rainfall; it is expressed in percentage [15] as:

$$R_{coefficient} = \frac{R}{P_e}\% \quad (2)$$

where R is the recharge and P_e is the effective rainfall.

3.4. Estimating Direct Runoff

Estimated runoff for water budget was developed by [16], and is given as:

$$R_{off} = 0.85 \times P - 30.5 \tag{3}$$

where R_{off} is the direct runoff and P is the precipitation.

3.5. Co-Integration Analysis

The climatic data was analyzed using co-integration analysis. The analysis involves a unit root test performed on levels, the first difference and the second difference were used to determine whether the individual input series are stationary and exhibit similar statistical properties. It must be noted that regressing a non-stationary time series data over another non-stationary time series data gives a spurious or unreliable regression. To correct for this, a unit root test is performed [17].

A time series is stationary when $X(t_1), X(t_2),..., X(t_n)$ is the same as the joint distribution of any set of $X(t_{1+k}), X(t_{2+k}),...,X(t_{n+k})$, for all n and k.

$$Y_t = pY_{t-1} + U_t \tag{4}$$

where $p \leq 1$ and U is the white noise error.

If the estimated p is statistically equal to 1 when Y_t is regressed on Y_{t-1}, then Y_t is non-stationary; that is, has no unit root (I (0)).

The Augmented Dickey–Fuller (ADF) test was used to test for the stationarity of the data. The test consists of the following regression:

$$\Delta Y_\tau = \beta_1 + \beta_{2\tau} + \delta Y_{\tau-1} \alpha \sum_{\tau=1}^{n} \Delta Y_\tau + \varepsilon_\tau \tag{5}$$

where:

$$\varepsilon_\tau = \Delta Y_{t-1} = (Y_{t-1} - Y_{t-2}), \Delta Y_{t-2} = (Y_{t-2} - Y_{t-3}) \tag{6}$$

The Johansen procedure was used to test for the number of co-integration vectors in the model. The Johansen technique was used not only because it is vector auto-regressive-based but because it performs better in multivariate models [18]. If $X_t$and Y_t are then co-integrated, their short-run dynamics can be described by an error correction model (ECM). The theory states that if two variables, Y and X, are co-integrated, then the relationship between them can be expressed as an ECM [18]. The co-integration model is given as:

$$P_t = \beta_0 + \beta_1 R_t + \beta_2 SR_t + \beta_4 MT_t + \beta_5 XT_t + \beta_6 ET_t + U_t \tag{7}$$

3.6. Correlation and Regression Analyses

Pearson correlation coefficient was used to evaluate the strength of the relationship between meteorological factors. The impact of independent variables (*i.e.*, humidity, temperature, rainfall, wind speed, and duration of daily solar radiation) on estimated recharge was evaluated using linear regression.

4. RESULTS

The methods discussed were used to estimate all parameters from January 1983 to December 2014. This covers a period of thirty-two (32) years. Table 1 shows the amount of calculated groundwater recharge, evapotranspiration, the recharge coefficient, and surface runoff. The ratio of recharge to total precipitation is 11%, which according to [16], is classified as a low rate of recharge. It is proved that aside from precipitation (with the high rate), other parameters (such as surface cover) also have a key role [4,5] in recharge. Table 1 shows the hydrological data of the study area for 32 years, for 1983 to 2014.

In Table 1, *P* is the annual precipitation, *R* is the annual groundwater recharge (mm), *ET* is the evapotranspiration (mm/year), *Min. T* is the minimum temperature (°C), *Max. T* is the maximum temperature (°C), and $R_{coeff.}$ is the recharge coefficient (%).

Figure 2 shows the annual rainfall, recharge, evapotranspiration, and temperature from 1983–2014. Groundwater recharge increases with precipitation and shows the same trend throughout the years. However, evapotranspiration shows variation in trend throughout the study area. This is expected because evapotranspiration is a function of temperature, which showed a remarkable change throughout the years, in the study area. Typically, low rainfall causes low groundwater recharge infiltrating into the aquifer, this is due to increase in temperature and evapotranspiration and *vice versa*.

Table 1. Hydrological data of the study area for 32 years (1983 to 2014).

Year	P (mm)	R (mm)	Min. Temp. (°C)	Max. Temp. (°C)	ET (mm/yr)	Run Off	R_{coeff}(%)
1983	901.6	159.0	22.3	32.1	1284.8	763.8	23
1984	1032.4	177.0	20.0	31.9	1252.0	874.9	20
1985	1185.7	196.0	21.5	31.9	1116.9	1005.2	21
1986	1210.7	199.0	24.8	31.9	1394.3	1026.5	20
1987	1224.5	199.4	26.4	37.7	1525.7	1038.2	21
1988	1576.1	237.8	20.9	36.3	1219.1	1337.0	19
1989	1222.6	200.3	19.4	37.1	2000.2	1036.6	20
1990	1336.1	213.0	20.4	37.6	2029.4	1133.1	20
1991	1185.0	195.9	20.1	36.5	1328.6	1004.7	20
1992	1445.6	191.2	18.5	37.4	1332.3	1226.2	20
1993	941.2	164.8	23.4	37.2	1357.8	797.4	22
1994	1009.6	174.2	31.4	37.3	1434.5	855.5	21
1995	1244.7	202.9	30.8	37.3	1401.6	1055.4	20
1996	1254.6	204.0	21.8	37.4	978.2	1063.8	20
1997	1233.5	201.7	29.9	37.4	1237.4	1045.9	20
1998	895.4	158.3	22.0	34.8	1076.8	758.5	22
1999	1037.5	177.9	13.8	31.9	981.9	879.2	23
2000	948.5	165.5	16.7	33.1	960.0	803.2	22
2001	950.0	165.5	15.3	33.3	1043.9	805.6	21
2002	1736.5	252.9	10.2	33.0	1284.9	1473.4	18
2003	1449.7	225.1	12.8	34.3	1325.0	1229.7	20
2004	1112.7	187.2	22.8	31.1	1262.9	943.2	20
2005	1112.7	187.2	13.4	33.0	1200.9	943.2	20
2006	1236.2	202.0	23.7	31.1	949.0	1048.2	21
2007	1943.9	273.1	26.0	32.2	1189.9	1651.4	18
2008	848.9	151.0	25.9	26.7	1215.5	719.0	22
2009	1189.9	197.8	26.0	30.5	1376.1	1008.7	20
2010	1233.9	201.7	25.8	36.4	1398.0	1046.2	21
2011	1497.7	223.5	25.4	36.6	1248.3	1245.0	15
2012	1241.5	204.2	25.0	36.8	1255.6	1052.7	15
2013	929.2	163.0	24.9	35.5	1379.7	787.2	22
2014	1058.0	180.3	22.9	32.2	1474.6	896.7	21

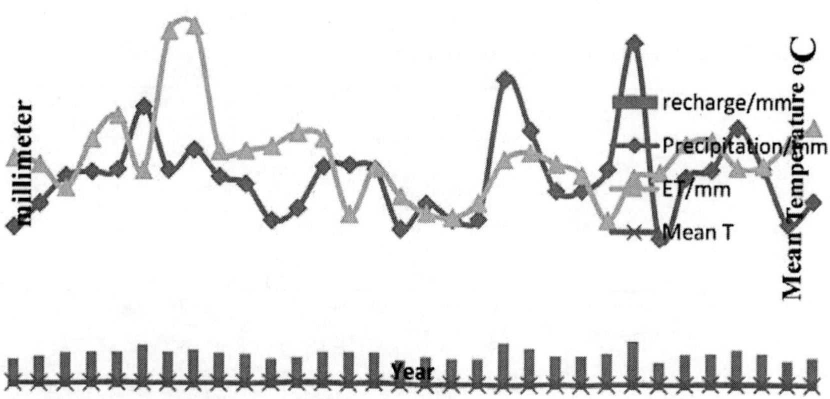

Figure 2. Total annual precipitation, groundwater recharge, evapotranspiration, and temperature.

There are objections to the use of either of these data since the data from the greater area is more useful for national decision-makers; the present method would be of great help to the local authority for future sustainable development of the region. The percentage of evapotranspiration, runoff, and recharge in the study area are 73%, 16%, and 11%, respectively. Typically low recharge causes high runoff over the surface.

4.1. Co-Integration

Table 2 represents the result of the ADF test. It is evident that recharge is non-stationary and likewise, the annual rainfall, but are stationary after the second differencing. Other climatic parameters are non-stationary. It has been ascertained that the variables exhibit unit root I (1) (non-stationary) at their levels, but stationarity after differencing, hence there exists a long run relationship between the climate parameters.

Table 2. Result of stationary test from the Augmented Dickey-Fuller Test.

Climate Parameters	Level	1 st Difference	2 nd Difference	Decision
Precipitation	−4.364246	−5.394125	−3.565695	Stationary
Recharge	−4.224929	−5.074373	−3.699143	Stationary
Min Temperature	−2.302965	−5.385516	−9.299585	Non-stationary
Max Temperature	−2.174135	−4.069986	−4.069986	Non-stationary
Evapotranspiration	−2.761985	−5.987190	−8.289337	Non-stationary
Relative Humidity	−2.604266	−4.506285	−6.032715	Non-stationary
Solar Radiation	−3.820166	−6.558539	−8.069424	Non-stationary

Critical level (1%): −3.6752.

Table 3. Johansen co-integration result.

Eigen Value	Likelihood	5% Critical	1% Critical	Hypothesized No. of CE(s)
0.751784	105.2439	94.15	103.18	None **
0.594706	64.83369	68.52	76.07	At most 1
0.481656	38.64253	47.21	54.46	At most 2
0.264670	19.58617	29.68	35.65	At most 3
0.223683	10.67054	15.41	20.04	At most 4
0.108415	3.327882	3.76	6.65	At most 5

** denotes rejection of the hypothesis at 5% (1%) significant level. Likelihood ratio test indicates 1 co–integration equation at 5% significant level.

Table 3 shows the Johansen co-integration results. The likelihood ratio shows that there is only one co-integrating (CI) equation in the analysis. The CI equation was based on the conformity of the coefficients with economic theory and its statistical significance. From the equation, the precipitation variable was considered to have a significant effect on estimated recharge in the study area during the period studied.

The adjusted R2 shows that about 98% of the variation in recharge is explained by the effects of precipitation (Table 4). This shows that an increase in rainfall value will cause the recharge value to increase and *vice versa*.

Table 4. Results from the Error Correction Model.

Variable	Coefficient	Standard. Error	t-Statistics	Probability
Recharge	11.13899	0.356742	31.22419	0.0000
Solar Radiation	−6.177148	7.792169	−0.792738	0.4354
Min Temperature	0.890065	3.595027	0.247582	0.8065
Max Temperature	2.818525	7.302545	0.385965	0.7028
Evapotranspiration	−12.55190	37.99302	−0.330374	0.7439
Constant	−935.5960	229.9052	−4.069486	0.0004

Dependent variable: precipitation. $R^2 = 0.977$; Durbin Watson stat: 1.583017; F-stat: 212.6064.

4.2. Correlation and Regression Analysis

This was to study the correlation between estimated recharge and climate parameters. Table 5 shows the Pearson correlation coefficient for meteorological parameters (1983–2014) on estimated recharge. The results showed that evapotranspiration was correlated with solar radiation at the 0.01 significance level, with $R^2 = 0.647$. Additionally, estimated recharge was correlated with precipitation at the 0.01 level ($R^2 = 0.943$). There is no significance at any level among the other parameters with regression analysis, which was conducted to evaluate the effect of climate parameters on estimated recharge. It was observed from Table 4 that all parameters showed minimal relationships to estimated recharge, except precipitation, which showed a dominant role of rainfall percolation of water into the ground in the study area. Table 5 andTable 6 show the Pearson correlation coefficient for meteorological parameters, and regression analysis of weather parameters on the estimated recharge in the study area.

Table 5. Pearson correlation coefficient for meteorological parameters.

	Precipitation (mm)	Recharge (mm)	Min Temp. (°C)	Max Temp (°C)	ET (mm)	Rel. Hum. (%)
Precipitation (mm)	1	0.943	0.279	0.258	0.059	−0.050
Recharge (mm)	0.943	1	0.433	0.231	0.064	0.051
Min. Temp. (°C)	0.104	0.085	1	0.261	0.327	−0.036
Max Temp. (°C)	0.244	0.263	0.205	1	0.019	0.303
ET (mm)	0.016	0.046	0.165	0.413	1	0.019
Rel. Hum. (%)	0.119	0.092	−0.118	0.389	−0.003	1
Solar Rad.	−0.013	0.023	−0.047	0.205	0.647	−0.017

Table 6. Regression analysis of weather parameters on estimated recharge.

Weathered Parameter / Recharge	Determination Coefficient (r^2)	Regression Equation
Precipitation (mm)	0.983	$Y = -962.462 + 11.2228X$
Min. Temperature (°C)	0.007	$Y = 20.2466 + 8.42E - 03X$
Max. Temperature (°C)	0.069	$Y = 31.5017 + 1.42E - 02X$
Relative Humidity (%)	0.008	$Y = 78.5326 + 1.22E - 0.2X$
Solar Radiation	0.001	$Y = 15.0630 + 1.39E - 0.3X$

5. CONCLUSIONS

The estimation of groundwater recharges have been carried out in Odeda Local Government Area, Ogun State, Nigeria. The methods used have greatly helped in estimating the groundwater recharges in the area. The annual mean groundwater recharge for 32 years (1983–2014) has been estimated to be 194.7 mm *per annum*. The mean of the recharge coefficient is 20.2%, the rate of evapotranspiration is about 73%, and groundwater recharge is about 11%. Therefore, 16% of rainfall water makes some allowance for runoff (and/or interflow) in the study area. The study shows that there is variability in climate parameters. The study also shows that the climate has a significant effect on groundwater resources. This is clearly revealed in the rainfall variable; however, evapotranspiration and solar radiation have a relationship with each other. The study also shows that the change in climate has a significant effect on recharge. This is clearly revealed in the rainfall variable; however, temperature seems to be an unimportant variable of climate in the determination of recharge in the study area. The correlation between parameters showed the highest positive correlation between recharge and rainfall. In addition, correlation between temperature and humidity, solar radiation, and ET are at the 0.01 level. Results of linear regressions proved precipitation has a significant effect (with $R^2 = 0.983$) on the estimation of recharge.

AUTHOR CONTRIBUTIONS

Victor Makinde conceived the research, Oluseyi Adeleke carried out the research which was thoroughly supervised by Victor Makinde and Ayobami Eruola. Oluwaseun Dada helped in acquiring the data. Oluseyi Adeleke, Akintayo Ojo and Taiwo Aluko analysed the data. Oluseyi Adeleke wrote the manuscript, Akintayo Ojo reviewed and adapted the manuscript to fit the style of the journal.

REFERENCES

1. Kishan, S.R.; Anil, K.M.; George, P.; Rakesh, K. Estimation of ground water recharge in Shankergarh block of Allahabad (India) using remote sensing and statistical approach. *Glob. J. Sci. Eng. Technol.* **2012**, *1*, 34–48.
2. Kumar, C.P. Estimation of natural ground water recharge. *ISH J. Hydraulic Eng.* **1977**, *3*, 61–74.
3. Bouwer, H. *Groundwater Hydrology*; McGraw-Hill Book Company: New York, NY, USA, 1978.
4. Shukla, S.; Jaber, F.H. Groundwater Recharge from Agricultural Areas in the Flatwoods Regionof South Florida. Available online: http://ufdcimages.uflib.ufl.edu/IR/00/00/15/42/00001/AE39900.pdf (accessed on 7 May 2015).
5. Sumioka, S.S.; Bauer, H.H. Estimating ground-water recharge from precipitation on Whidbey and Camano Islands, Island County, Washington, water years 1998 and 1999. *Water-Resources Investigations Report, 03-4101*; Tacoma, W.A., Ed.; U.S. Geological Survey: Pembroke, MA, USA, 2003.

6. Xi, C.; Zhang, Z.; Zhang, X.; Chen, Y.; Qian, M.; Peng, S. Estimation of groundwater recharge from precipitation and evapotranspiration by Lysimeter measurement and soil moisture model. *J. Hydrol. Eng.* **2008**, *13*, 333–340.
7. Kumar, C.P.; Seethapathi, P.V. Assessment of natural ground water recharge in upper Ganga Canal command area. *J. Appl. Hydrol.* **2002**, *15*, 13–20.
8. Oke, M.O.; Martins, O.; Idowu, O.; Aiyelokun, O. Comparative analysis of empirical formulae used in groundwater recharge in Ogun-Oshun River Basins. *J. Sci. Res. Repo.* **2013**, *2*, 692–710.
9. Chaturvedi, R.S. A note on the investigation of ground water resources in western districts of Uttar Pradesh. In*Annual Report*; U.P. Irrigation Research Institute: Bahadrabad, India, 1973; pp. 86–122.
10. Kumar, C.P. *Groundwater Assessment methodology*; National Institute of Hydrology: Roorkee, India, 2009.
11. Saghravani, S.R.; Ismail, Y.; Sa'ari, M.; Saghravani, S.F. Estimating groundwater recharge using empirical method: A case study in the tropical zone. *Sains Malaysiana* **2013**, *42*, 553–560.
12. Risser, D.W.; Gburek, W.J.; Folmar, G.J. Comparison of recharge estimates at a small watershed in east-central Pennsylvania, USA. *Hydrogeol. J. 2008,* **2008**, *17*, 287–298.
13. Ako, B.D. Geophysical prospecting for groundwater in parts of south-western Nigeria. Ph.D. Thesis, Department of Geology, University of Ife, Ile-Ife, Nigeria, 1979.
14. Allen, R.; Pereira, L.A.; Raes, D.; Smith, M. Crop evapotranspiration. In *FAO Irrigation and Drainage Paper No. 56*; FAO: Rome, Italy, 1998.
15. Misstear, B.D.R.; Brown, L.; Daly, D. A methodology for making initial estimates of groundwater recharge from groundwater vulnerability mapping. *Hydrogeol. J.* **2009**, *17*, 275–285.
16. Muteraja, K.N. *Applied Hydrology*; Tata MGraw-Hill: New Delhi, India, 1986.
17. Ayinde, E.O.; Muchie, M.; Olatunji, G.B. Effect of climate change on Agricultural productivity in Nigeria: A co-integration model approach. *J. Hum. Ecol.* **2011**, *35*, 189–194.
18. Maddala, G.S. *Introduction to Econometrics*, 3rd Ed. ed; John Wiley and Sons Ltd: New York, NY, USA, 2001.

6

CHAPTER

RISK MANAGEMENT AT THE LATIN AMERICAN OBSERVATORY

Á.G. Muñoz[1, 2, 3], D. Ruiz[1, 3, 4], P. Ramírez[1, 5], G. León[1,], J. Quintana[1, 7], A. Bonilla[1, 5], W. Torres[1, 2], M. Pastén[1, 8] and O. Sánchez[1, 9]

[1] Observatorio Latinoamericano de Eventos Extraordinarios (OLE2),
[2] Centro de Modelado Científico (CMC). Universidad del Zulia, Venezuela
[3] International Research Institute for Climate and Society (IRI). Columbia University in the City of New York, USA
[4] Escuela de Ingeniería de Antioquia, Colombia
[5] Comité Regional de Recursos Hídricos (CRRH), Costa Rica
[6] Instituto de Hidrología, Meteorología y Estudios Ambientales (IDEAM), Colombia
[7] Dirección Meteorológica de Chile (DMC), Chile
[8] Dirección Nacional de Aeronáutica Civil (DINAC), Paraguay
[9] Servicio Nacional de Meteorología e Hidrología (SENAMHI), Perú

1. INTRODUCTION

The *Observatorio Latinoamericano de Eventos Extraordinarios* (OLE^2), or *Latin American Observatory*, is a regional collaborative network that, in addition to the existing infrastructure in each country, aims to ultimately increase the efficiency of the decision-making processes, especially in terms of getting more accurate environmental information and exchanging experiences on data, methodologies and scientific products, all of which are done with a standardized methodology and a Web-sharing service.

Present partners of this collaboration are national weather services, universities and research institutes of Central America, the Andean countries and from Southeast South America (see Figure 1). Nowadays, the Centro de Modelado Científico (CMC) of Universidad del Zulia, the founder and coordinator of the initiative, regionally coordinates the OLE². The coordinating role involves, among other tasks, the suggestion of (a) methodologies for the provision of scientifically based tools and products, (b) mechanisms for the successful use of the environmental information in the policy making process at the different levels (e.g. national and province governments, private sector), (c) facilitation of continuous and effective communication of the different partners and (d) the interchange of experiences and products (by means of the Observatory's products web interface[1] wiki page[2], forum[3] and email list[4]). For more information see references [1] and [2].

The goal of the OLE² is to monitor and forecast key environmental variables and develop accurate products based on different scientific tools in order to help decision makers improve risk management and set up efficient early warning systems. The Observatory provides several model outputs for meteorological, seasonal, and hydrological forecasts, 5-day high-resolution oceanographic prediction for the eastern Pacific, droughts, fire and flood indices, ecosystem dynamics (like duckweed/algae occurrence in Lake Maracaibo), and climate and health applications(e.g., regarding malaria), among others.

The Observatory is organized in interconnected working groups (WGs) or axes: namely, meteorology; climatology; hydrology; air quality; climate and health; and climate change and variability. Each WG has its own methodologies and products, but all of them exchange information and results with other axes and countries if necessary. It is also regionally divided in three branches: the Observatorio Andino, Observatorio Centroamericano and Observatorio del Sudeste de Sudamérica (Andean, Central American and the South Eastern South America Observatories). They operate using the infrastructure already in existence in the different regions by consolidating the collaboration networks through the recognition of common needs and the provision of interdependent solutions. The final products, be it a seasonal forecast or an early warning system for the last-mile user, is built between different partners by sharing their experiences and expertise on different fields. This has created a strong identification between the participating institutions and the Observatory's products. This fact and the continuous need of collaboration to provide regionally successful decision-making tools on time, have proven a natural way to guarantee the sustainability of the initiative, originated in 2007 in Venezuela as a national observatory.

Advances and difficulties are shared and discussed through e-mail lists and videoconferences. The purpose of the videoconferences is to discuss the ongoing projects, products, and new methodologies, and to provide special assistance on models and several other technical issues.

The Observatory has also provided a Grid Technology Infrastructure since 2010, known as AndesGrid [3] to help institutions with less computational power to increase their capabilities using other institutions' available computer resources. The initiative enables the partners to share not only model outputs and, but also their local observations (e.g. rain gauge and temperatures) in "real time." For additional details, see [3].

FIGURE 1. The Observatorio Latinoamericano's partners (as January 2012) are mainly national weather services, universities and research and development institutes. (The red line links the countries participating in the AndesGrid iniciative [3]. See main text).

Of the OLE²'s main achievements in this period, the most important is the enhancement of the institutions' human resources by means of continuous technical support for forecast, modeling, verification, risk assessment and management and many other issues, all of which is done in their native language. The project has also succeeded in standardizing forecasts, data formats, and methodologies, providing common models, tools, and procedures that are used on a daily basis in all the involved institutions. A key tool has been the wiki, which contains all of the needed steps for each task, which by itself is another of the most important elements for ensuring the long-term continuity of the OLE².

Among the different efforts, risk assessment and management are some of the most important tasks targeted by the Latin American Observatory partners. Even when a general framework is used by the OLE² on this field, the different institutions develop tailored applications to take advantage of the local experience and mechanisms already in existence in their countries. However, the general framework closely follows the ideas discussed by Mora [4].

Risk management is from a wide perspective, an approach to the issue of risk and disasters and from a more institutional perspective, a policy system that, among its components, permits stakeholders in several levels, to identify, analyze and quantify the probability of human, material and ecosystem losses, as well as to provide metrics and technical criteria for prevention, mitigation, reduction or transfer of risk [4]. A key idea to be still implemented in most parts of the world is that risk management needs to be understood and addressed as an investment, not as a cost. [4] It is clear that nowadays the establishment of efficient risk reduction plans does not possess a real priority on the agenda of decision-makers and stakeholders, except *a posteriori*. This is especially true in most parts of Latin America, as is exposed in some case studies included this chapter. Moreover, in general, there are no consolidated cultures incorporating risk management throughout decision-making processes for public and private investment and planning (see [4] and references therein). Also, private sector involvement in this process presents several unique challenges to be overcome [5].

Since 1975 in Latin American and the Caribbean (LAC), the costs of losses attributable to major natural hazards have been estimated at around US$300 billion, with more than 280,000 human deaths and affecting more than 160 million people [6-8]. Table 1 references the major natural disasters in LAC in the last three decades.

TABLE 1. Major natural disasters in LAC between 1983 and 2011 [6-8].

Year	Hazard/Event	Countries
1983	Floods	Argentina, Bolivia, Brazil, Peru
1983	Earthquake	Chile, Colombia
1985	Earthquake	Mexico
1998	Floods, landslides associated to hurricanes George and Mitch	Central America and the Caribbean
1999	Floods and landslides	Venezuela (Vargas State)
2005	Most active hurricane season recorded in history [8]	Central America and the Caribbean
2009	Drought (strong El Niño event)	Venezuela, Colombia and Ecuador
2010	Earthquake	Chile
2010, 2011	Floods (strong La Niña event)	Panama, Colombia and Venezuela

Consequently, it has been an urgent need for the Latin American Observatory's partners to address the establishment of effective policies for risk assessment and management in their respective countries. To this end, an implementation plan was suggested by CMC and this chapter presents briefly the adopted definitions and general ideas (sections 2-4), as well as case studies for different managerial sectors in Latin America (section 5) in order to discuss in some detail, present risk assessment/management policies and methodologies. Finally, we discuss the benefits of sharing experiences among different countries to address common problems in geographically complex regions, as well as how to achieve success in countries where the scarcity of economic and trained human resources imposes serious limitations on the effectiveness of risk management systems.

2. RISK DEFINITION AND MANAGEMENT

In this and the next section the main definitions related to risk management, hazards and vulnerabilities are discussed in order to provide the foundations chosen by the Observatory in these matters. We don't pretend to address all the details in this complex field, but to outline the main ideas for future reference.

Different agents play a role in the occurrence of natural hazards, which derive from the damaging potential or a combination of them. These agents are frequently [4,6] classified in terms of their origin, namely

- hydro-meteorology: both global and local processes (hurricanes, floods/droughts, weather extreme events)
- internal geodynamics (seismicity and volcanism)
- external geodynamics (landslides, intensive erosion, torrential debris flows).

Following Mora [4], natural hazards (*H*) can be formally defined in terms of the probability that an event becomes so intense (*a*) within time and space frames that it produces significant damage (*d*). The intensity and damage definitions depend on the region and timescale under consideration. Vulnerability (*V*) can be associated with the probability that, according to the intensity of the natural event, damage might occur as a function of the degrees of exposure and fragility of the elements involved. Risk (*R*) therefore is the combined probability (convolution, *) that a hazard might cause significant damage, according to the following relationship:

$$\int_{h,d} p(R)da = \int_{h} p(H)da * \int_{d} p(V)da \tag{1}$$

The probabilities are not, in general, constant, so the spatial and temporal variability must always be carefully considered. Moreover, *V* and *H* are not independent.

Risk management involves not only the quantification of the probability of loss via equation (1) and the derived secondary effects, but also the generation and execution of policies for [4]

- identification, analysis and quantification of risk
- formulation and application of measures of prevention, reduction and mitigation of causes
- financial, social and environmental protection
- execution of related protocols (e.g. preparedness, response, rehabilitation, reconstruction).

Efforts to foster these processes may include the establishment of adequate policies and protocols to incorporate stakeholders into risk management, improvement of the climate/geodynamic information available, identification of the different sources of vulnerability and their evolution in time, definition of "acceptable" risk levels and the establishment of early warning and monitor systems, and the adequate ways to communicate their products to the end-mile users.

Risk management policies foster and integrate strategically balanced processes capable of providing decision-making tools in a synergistic, decentralized and participative way. In the OLE² this is a central philosophy, which also enables the treatment of risk in trans-boundary zones (e.g. the Observatory enabled the continuous exchange of climate services between the National Weather Services and decision makers during the 2010 floods that affected Panama, Colombia and Venezuela in similar ways).

An important aspect of risk management is related to operative platforms that enable decision makers and users to interact and to obtain information useful to assess vulnerabilities, the impact of natural hazards and the risk probabilities for a certain region. This infrastructure is aimed at providing this kind of information *ex ante*, but it can also be used during or after an emergency. It usually includes

- macro- and micro-zoning of hazards and vulnerability and its probable evolution in time (see section 4)
- damage assessment
- identification of priority areas of intervention
- protocols for vulnerability reduction (hazards are not, in general, reducible)
- instruments for financial strategy and budgetary allocations
- identification of "risk bearers" and their responsibilities
- administrative and legal information (e.g. design, operation, lease, transfer and reclamation contracts)
- information on past events
- areas of further investments to reduce future disasters.

The implementation of these platforms in LAC is more advanced in a few countries (see section 5.1 for a Central American example). The Observatory's partners are taking advantage of the experience developed by other members of the regional initiative to improve their own risk management infrastructure.

Finally, but no less importantly, the Observatory is now committed to following the International Research Institute on Climate and Society's (IRI) Four Pillars for Climate Risk Management [31], namely

1. **Identify vulnerabilities and potential opportunities** due to climate variability/change for a given water, agriculture, or health system.
2. **Quantify uncertainties in "climate information"** in order to reduce uncertainties in using that information.
3. **Identify technologies and practices** that optimize results in normal or favorable years as well as technologies and practices that reduce vulnerabilities.
4. **Identify interventions, institutional arrangements and best practices** that reduce exposure to climate vulnerabilities and enable the opportunistic exploitation of favorable climate conditions.

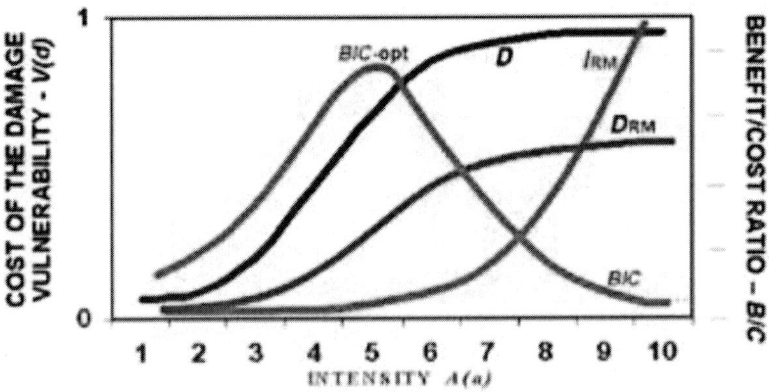

FIGURE 2. Benefit/cost (B/C) ratio defined as a function of damage expected without risk management (D), reduced by applying risk management (D_{RM}) and the investment involved (I_{RM}). (After Mora [4])

3. BENEFITS OF RISK MANAGEMENT

It is possible to use risk management as a tool for providing an idea of the economic value of potential damages caused by hazards and the actions undertaken to reduce vulnerabilities. Figure 2 sketches an accumulating damage curve *D* without risk management. If risk management policies are used, this same curve is reduced to a more optimistic distribution, *DRM*. The reduction of damage due to risk management is then

$$RD = D - D_{RM} \tag{2}$$

The associated investment (*IRM*) for such a reduction depends on both the hazard and vulnerability involved (which, again, are not independent). On the other hand, the Benefit/Cost (*B/C*) ratio allows for the finding of optimal levels of investment. It's possible to show [4] that the net benefit of risk management is

$$B_{RM} = (D - D_{RM}) - I_{RM} \tag{3}$$

and the Benefit/Cost ratio can be defined as

$$B / C = \frac{D - D_{RM}}{I_{RM}} - 1 \tag{4}$$

Having real distributions of the different quantities it might be possible to select optimal points (e.g. the saddle point named *B/COPT* in Figure 2) to limit profitable investments. See [4] and [6] for details.

4. A MULTI-SCALE APPROACH FOR RISK MANAGEMENT

The traditional approach to computing risk probabilities using equation (1) involves the consideration of static or, in the best-case scenario, long-term mean vulnerability maps of the associated hazards. Nonetheless, the probability density function associated with hazards in equation (1) is frequently computed from time series that may contain useful information at different time scales. Given the fact that a decision maker might be interested in a certain time scale (e.g. on the order of its own managerial period), a temporal decomposition of the original time series used to identify the hazard probability may be extremely helpful.

For example, Figure 3 presents such time-scale decomposition for Southeast South America's precipitation. The lower panels sketch the non-linear long-term trend (a proxy for climate change), and decadal and interannual variability time series [9]. The rationale is that in considering the interaction of the three climate signals (the explained variance being a way to ponder the specific weight of each one of them) it is possible to provide better climate services to decision-makers, that are more adequate for their scale of interest, e.g. next 10 years.

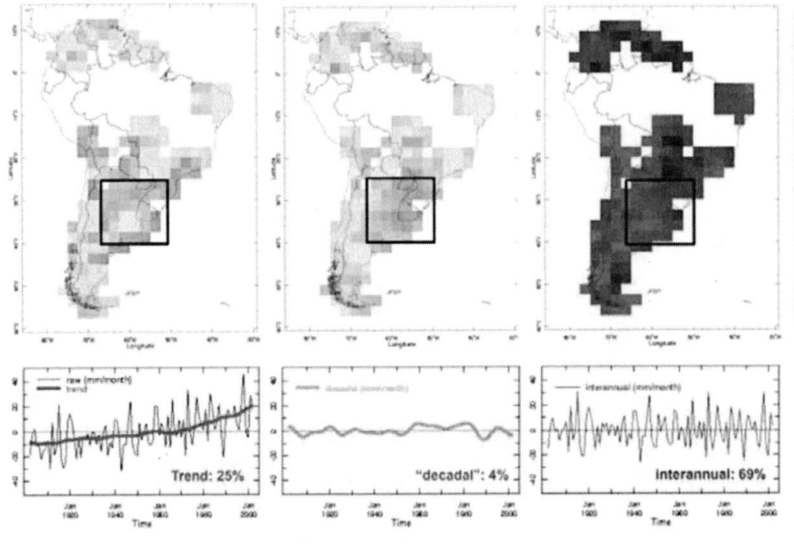

FIGURE 3. Decomposition for Southeast South America's precipitation (boxed region) for three different time scales: long-term trend (left panel), decadal variability (central panel) and interannual variability (right panel), with their respective explained variance (colors in the map and values in the bottom corner of each time series). Grid boxes in white are not statistically significant at a p-level=0.95. (After Goddard *et al.* [9])

As mentioned before, hazards and vulnerabilities are not independent. The latter encompasses exposure to hazards, sensitivity to these hazards and adaptive capacity. All of them, in fact, evolve in time. Therefore, a similar multi-scale time decomposition must also be done to correctly assess the related vulnerabilities. This provides a more realistic risk probability distribution, and thus better assessments and policies can be developed.

5. CASES OF STUDY

In this section a few state-of-the-art examples of risk assessment and management in several countries in Latin America are briefly described. Signatory and participant institutions of the Latin American Observatory operate over these structures, providing tailored methodologies based on what has been explained in previous sections.

5.1. Seasonal Climate Outlooks Applications for Food Security Decision-Making in Central America

In Central America, seasonal climate outlooks turn into risk scenarios used by food-related sectors to help support their decisions and prevent food insecurity. This is a coordinated effort with specialized entities of the Central American Integration System.

The Central American Climate Outlook Forum (CA-COF), coordinated by the Comité Regional de Recursos Hidráulicos (CRRH-SICA)[5] - (Regional Water Resources Committee), has consolidated a process to issue three seasonal outlooks per year, bringing together the capacities of all seven weather services in the region. The CA-COF takes advantage of international and global sources of information, which it analyzes along with its own sources and historical data to produce Climate Outlooks for the Central American region. Over the last decade the Forum has issued 38 regional climate outlooks.

To facilitate the use of climate risk information for decision-making with the aim of reducing food insecurity risks, CRRH-SICA has fostered a mechanism over the past few years that turns the Outlook into risk scenarios for those sectors related to nutrition and food security, particularly agriculture, fisheries, potable water and public health, as well as cross-cutting areas like risk management and emergency response. These scenarios are used when deciding preventive measures to mitigate the impact of climate variability on food security.

In a joint effort between the specialized agencies of the Central American Integration System (SICA) and the Regional Food Security Program (PRESANCA II) funded by the European Union, immediately after an Outlook is issued, a group of regional experts is convened to enhance the Outlook with information from the different sectors mentioned above, turning the product into sectoral climate risk scenarios that could guide early warnings of actual and potential hazards to food security.

The working group, composed of CA-COF members and the experts, uses these scenarios to identify sector-specific preventive measures. The food security risk scenarios and the suggested measures are circulated among government entities and other organizations involved in addressing food and nutritional security through their own networks to ensure all this information reaches the most appropriate decision makers.

This same method is used at the national level to produce specific recommendations for diverse sectors in each individual country. Those are then distributed among national authorities and stakeholders in private sector with the seasonal outlook map and lists of suggested prevention measures. An example of the matrix of climate risk for agriculture is presented in Table 2.

TABLE 2. Central American Climate Outlook Forum risk assessment for two crops (maize and beans), for December 2011-February 2012.

Climate Risk /potential damages for Main Crops. Quarter December 2011 -March 2012							
Crop				Country			
	Belice	Costa Rica	El Salvador	Guatemala	Honduras	Nicaragua	Panama
Maize	Plantations in coastal areas likely to be affected by above-normal preci-pitation	No risk	No risk	Risk of post-harvest damages due to higher humidity associated with above-normal rainfall. Increased cost of post-crop product manage-ment	Risk in the Cari-bbean planting area due to above-normal rainfall. Second planting season likely to be affected	Second crop in the Atlantic Autonomo us Regions likely to be affected by above-normal rainfall. Exportation s likely to be impacted	Crop loss risk in Western Cari-bbean area. Posible impacts in food security of indi-genous population.
Beans	Plantations in coastal area likely to be affected by above-normal precipitation	Risk of crop damage for plantations in Northern Plains, Cari-bbean area and South Pacific	No risk	No risk	Risk in the Cari-bbean planting area due to above-normal rainfall. Second planting season likely to be affected	Second crop in the Atlantic Autonomo us Regions likely to be affected by above-normal rainfall. Yield reduction may reach 15% .	Loss of crop risk in Western Caribbean area. Posible impacts on food security of indigenous popu-lation.

As part of the same, routine-specific measures are identified at a national level by similar working groups of climatologists and national experts, and suggested to authorities.

A process is now underway to evaluate the use of outlooks and climate risk information directly with the beneficiaries, both for the private and the public sectors, to strengthen decision-making efforts.

5.2. A Flood Early Warning System for Vargas, Venezuela

In Venezuela, heavy rains represent a significant problem for human populations, not only in rural areas where farms and crops are affected, but also in some urban settings where many inhabitants reside in poorly-constructed houses that are highly vulnerable to floods and landslides, or that are sometimes located on steep terrains and floodplains.

FIGURE 4. Aerial view of Caraballeda, Vargas state (Venezuela), after mudslides and debris flows event on December 15th 1999.

On December 15th, 1999, on the northern coast of Venezuela, torrential rains led to flash floods and debris flows that killed tens of thousands of people, destroyed thousands of households, and meant the complete collapse of the area's infrastructure (see Figure 4). The "Vargas disaster", as it has been known ever since,, is considered the worst natural disaster in Venezuela's last half century (Table 1). Even though flood prediction is an essential piece of Climate Risk Management, Venezuela did not have, at the time, a consolidated early warning system that could alert decision-makers and stakeholders about this extreme event. In this section, the implementation of an early warning flood system for Vargas state is described in some detail. It is a completely general methodology, so other regions in LAC can benefit from the experience described here.

The Vargas disaster is also an important case study on flood prediction, not only because of its unusual rainfall amounts, but also for the nature of the terrain where it took place. The basin is located in a mountainous region of metamorphic rock, coarse soils and steep slopes, making the area highly vulnerable to floods and debris flows [10].

The early warning system involves the estimation of vulnerabilities and probabilities related to heavy precipitations and mudslides. Guenni *et al.* [11] has computed the vulnerabilities using the total affected and exposed people, considering both the spatial and temporal variability, as discussed in section 4. As an example, Figure 5 presents a zoom-in for most parts of Venezuela showing months of rainfall exposure in colors and the population density as black pixels (for details see [11]). Regions near the Venezuelan coast are in general more vulnerable (due to more population density and exposure). Using map algebra in a Geographical Information System (GIS), the map is then intersected with other maps containing information about the terrain (e.g. water holding capacity or mudslide probabilities). Following equation (1), the final map, in sketching homogeneous vulnerability zones, must be written in terms of probabilities for different hazard intensities.

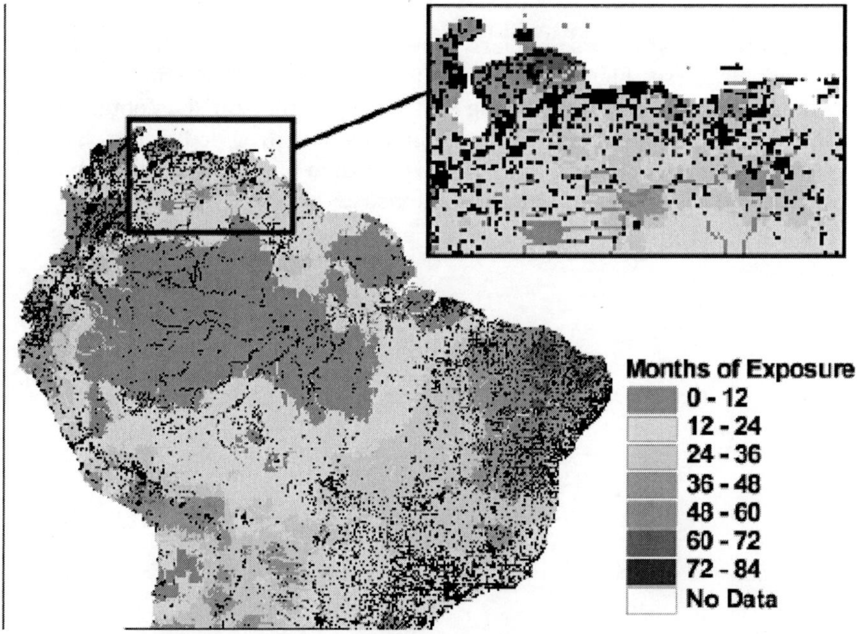

FIGURE 5. Map of exposition (in colors) and population density (pixels in black) for Venezuela. After Guenni*et al.* [11]

On the other hand, in order to compute flood risk probabilities, hazard probability maps must be produced (also in terms of intensities). In this case, Torres and Muñoz [12,13] have suggested a methodology based on hindcasts involving an off-line coupling of a regional climate model, a process-based hydrological model and a routing model.

The Climate Weather Research and Forecasting model (CWRF) was used to simulate 13 years (1996-2008) of data, paying special attention to the Vargas disaster atmospheric conditions. Rainfall and maximum and minimum temperatures were bias corrected using observed values. Then they were used as input data along with soil properties and vegetation parameters for the Variable Infiltration Capacity model (VIC). This energy and mass balance model was finally coupled to the Lohmann routing model to produce realistic stream flows that adequately considered the local topography (for details about the different models and procedures see [2,12,13] and references therein). Simulation outputs were consistent with the flooding event of December 1999. The study was able to reproduce how long rains, along with low evapo-transpiration due to high cloudiness, contributed to saturate soil layers. Moreover, the study was able to identify critical values (see Figure 6) in order to establish "yellow" and "red" alerts in the Vargas flood early warning system, using the statistics obtained for the 13 years simulation and local stream flow data provided by the National Weather Service. Model outputs were then used to create hazard probability maps in terms of hazard intensity.

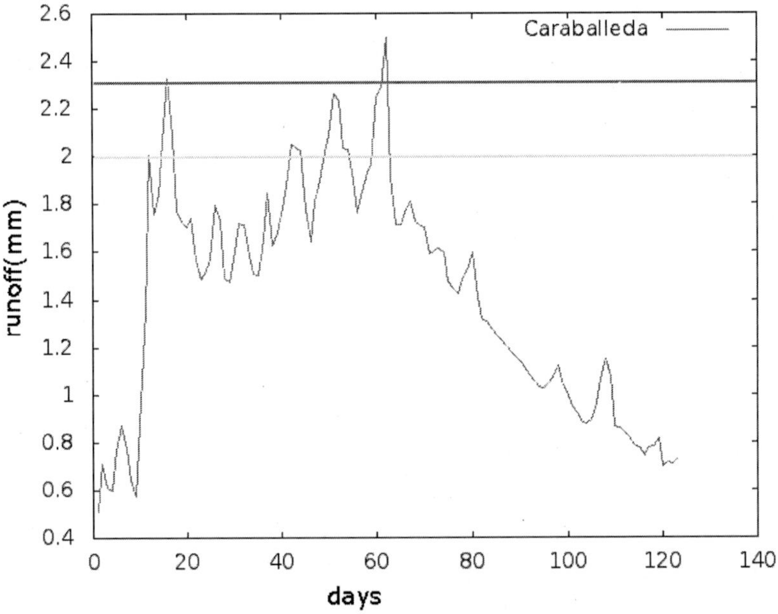

FIGURE 6. Figure 6: Caraballeda's simulated runoff time series for the Vargas disaster event using CWRF, VIC and the Lohman routing model (see main text). The runoff has been normalized using the standard deviation. The critical day (December 19th, 1999) corresponds to simulation day 62 in this plot. The yellow and red lines at 2σ and 2.3σ, respectively, were defined as the "yellow" and "red" alerts in the designed early warning system. (After Torres and Muñoz [12,13])

As a final result, series of maps are provided with the hazard's probability of occurrence in terms of the hazard intensity. They are produced computing the convolution between the corresponding hazard and vulnerability maps, following equation (1). For additional details see [12,13].

This probabilistic approach has an important advantage: it takes into consideration the possible uncertainties related to each one of the processes involved in the final flood risk probability map.

5.3. Bridging the Gap Between Climate Information and Health Services in Colombia and Ecuador

Climatic factors play a significant role in the transmission dynamics of several infectious diseases [14]. Therefore it should be a critical priority to incorporate climate information into disease risk assessments and early warning and response systems. (See figure 7.) This, in fact, has been one of the goals of the Colombian and Ecuadorian meteorological and health services in recent years. In Colombia, in particular, the National Institute of Health (INS) recently teamed up with the Institute of Hydrology, Meteorology and Environmental Studies (IDEAM) and several universities and research centers to design and implement a proactive, collaborative, multidisciplinary, Integrated Surveillance and Control System - ISCS [15,16]. This initiative is part of a set of measures and policy options that the Colombian adaptation strategy to climate change proposed for three areas of primary concern: the high-altitude Andean ecosystems, the insular and coastal areas, and human health (see Integrated National Adaptation Pilot [16]).

The aim of the adaptation strategy for the health sector is to have a better-prepared institutional arrangement for increased exposure to malaria and dengue fever, two climate-sensitive, vector-borne diseases that are still considered human health burdens in Colombia [17]. According to the scientific literature, the potential increase in the incidence of these two diseases is likely to occur not only in the already endemic malaria and dengue fever prone areas, but also on the fringes with the Andean regions where local communities have not been exposed to pathogens before. Thus they lack the immunity against these microorganisms. The approach has been to assist the health sector to better cope with current climate variability and climate-related events, as a means to make it better prepared against future climatic conditions likely to be brought by the ongoing long-term global climate change. As a result, Colombia has been working on reducing people's vulnerabilities to the negative impacts of malaria and dengue outbreaks, as well as developing an Early Warning System framework, supported by seasonal forecasting capabilities, weather and environmental monitoring, and statistical and dynamic models [18-23].

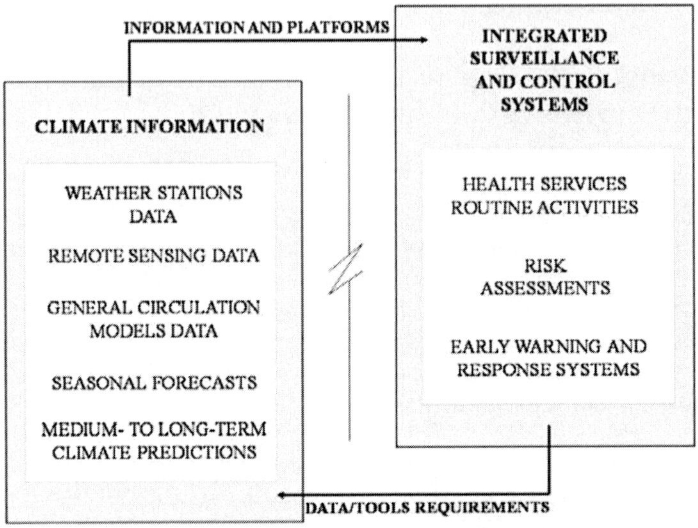

FIGURE 7. Climate information and human health integrated surveillance and control systems.

This effort required linking what had previously been two largely separate analytical domains: the field of public health, traditionally dominated by human health experts and practitioners; and weather and climate science, usually lead by meteorologists, climatologists, engineers, and environmental science experts. Several biologists, entomologists and social experts have also joined this collaborative, inter-institutional and interdisciplinary effort. The implementation of the ISCS has required, among many other activities, analyzing the local eco-epidemiological settings of various malaria- and dengue fever-prone pilot sites, implementing process-based biological and statistical models, and strengthening the local capacity of health authorities. It has also required strengthening the IDEAM capability to routinely and systemically produce and disseminate, relevant, continuous, homogenous, and reliable climate information that could support the decision-making process of health authorities. Such information includes ground-truth historical records, modeling simulation outputs, seasonal forecasts, El Niño Southern Oscillation forecasts, and climate change predictions. All this information, along with disease morbidity profiles, entomological conditions, socioeconomic drivers, and impacts of interventions and control campaigns, is now steadily becoming a core part of routine activities and disease control plans of health services at regional, municipal and local levels, and is starting to facilitate a better allocation of health resources and more cost-effective preventive responses. On a broader scale, measures proposed as part of the adaptation plan have been mainstreamed into the Colombian political agenda to ensure their sustainability. They have reached, for instance, the 2010-2014 National Development Plan, the 2010-2014 Environmental Action Plan, the Colombia's Poverty Reduction Plan, and the Public Health National Plan.

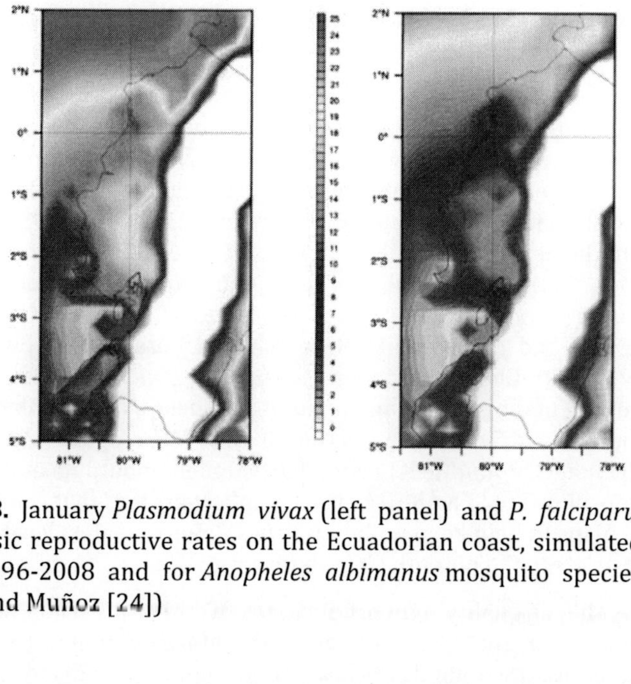

FIGURE 8. January *Plasmodium vivax* (left panel) and *P. falciparum* (right panel) basic reproductive rates on the Ecuadorian coast, simulated for the period 1996-2008 and for *Anopheles albimanus* mosquito species. (After Recaldo and Muñoz [24])

In Ecuador, in turn, the National Institute of Meteorology and Hydrology (INAMHI) has conducted various research activities [24,25] on the analysis of potential increases in the incidence of malaria and dengue fever in key lowland provinces. Activities have included the implementation and coupled analysis of the 30 km WRF (Weather Research and Forecasting), dynamic downscaling regional climate model, and the Ross-Macdonald malaria infectious disease process-based model, to reproduce the spatio-temporal variability of primary positive cases reported by the National Malaria Eradication Service over a recent 13-year historical period (see figure 8). They have also included a first design of an Early Warning

System for malaria and dengue fever outbreaks, which is now developing into an integrated surveillance and climate modeling for malaria and dengue fever predictability in rural and urban settings. The INAMHI has also joined efforts with several research groups to broaden the understanding of the complex transmission dynamics (in both space and time) of these vector-borne diseases, in order to improve decision-making processes in regional and local health authorities and, in a more general sense, to strengthen the already solid surveillance and control activities of infectious diseases conducted by the Ecuadorian health sector.

5.4. Air Quality Risk Assessment For Lima And El Callao (PERU)

Poor air quality conditions can cause many respiratory problems including asthma and severe allergies and can also pose serious health problems such as cancer. In order to issue public warnings of high pollution episodes, the Peruvian National Service of Meteorology and Hydrology - SENAMHI is working on the design and implementation of an air quality forecast system for the cities of Lima and El Callao. The system maps out current and future ambient air quality conditions based on hindcast simulation runs of the BRAMS model and the Weather Research and Forecasting coupled with Chemistry (WRF-Chem) model (see figure 9), as well as on *in situ* measurements of main air pollutants, such as particulate matter smaller than 10 and 5 micrometers (PM10 and PM5, respectively), nitrogen oxides (NOx) and ground-level ozone (O_3). Modeling outputs and real-time information allow the SENAMHI to issue public warnings that could activate municipal plans to help keep pollution levels down and alert local residents in Lima districts to the potential health threats.

Up to date, the air quality network includes five high-precision monitoring stations located in the surroundings of the aforementioned two densely populated cities. (See figure below.) High pollution episode warnings include events such as the one that took place in the localities of Ate and Villa María del Triunfo on January 1, 2012. *In situ* air quality measurements (figure 10) suggested that PM10 concentrations reached 295 and 290 $\mu g/m^3$, respectively, which exceeded in 45 and 40 $\mu g/m^3$ the 250 $\mu g/m^3$ daily concentration threshold. In the monitoring station Santa Anita, in turn, the PM10 concentration reached 199 $\mu g/m^3$, also exceeding the health-based daily air quality standard concentration of 150 $\mu g/m^3$. The air quality stations San Borja and Campo de Marte reported moderate daily concentrations of about 78 and 54 $\mu g/m^3$, respectively.

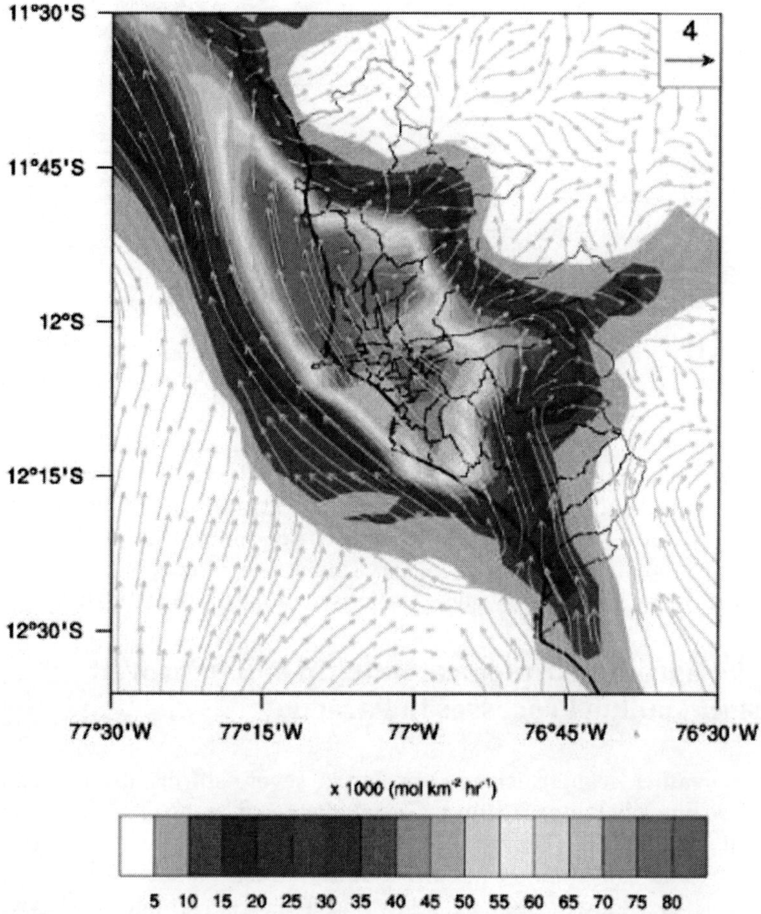

FIGURE 9. 4-km spatial resolution hindcast WRF-Chem model simulation outputs of NOx concentration fluxes in the geographic domain 12°30'S – 11°30'S and 76°30'W – 77°30'W. Typical NOx concentration fluxes are expressed in thousand mol/km^2/hr. The reference arrow represents wind speeds of 4 m/s.

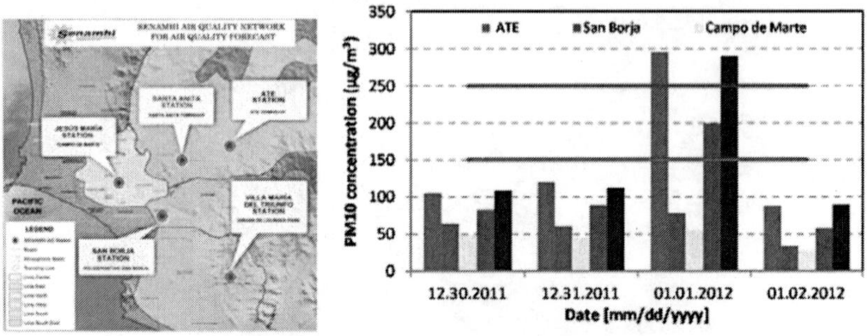

FIGURE 10. 10 micrometer or less particulate matter (PM10) concentrations (right panel) gathered at the SENAMHI air-quality monitoring stations Ate, San Borja, Campo de Marte, Santa Anita, and Villa Maria del Triunfo (see locations in left panel) over the period spanning from December 30, 2011 through January 2, 2012. The red solid line depicts the health-based daily air quality standard concentration of 150 μg/m^3. The violet line depicts, in turn, the 250 μg/m^3 daily concentration threshold above which high pollution episode warnings are issued. Air quality is considered 'good' when daily concentrations do not exceed 50 μg/m^3.

5.5. The Agrometeorological Bulletin and Improved Decision-Making Processes in Paraguay

Extreme weather events such as river floods, severe storms, droughts and below-freezing low temperatures are strongly linked to the onset of El Niño Southern Oscillation (ENSO). Extreme events primarily affect river streamflow and cause numerous direct and indirect impacts on many key sectors of the Paraguayan economy: agriculture, ground and river transportation, potable water, construction, electricity, and recreation. Several studies have demonstrated that the potential impacts of these extreme events and their economic consequences are directly related to the magnitude of ENSO episodes. Historically, Paraguay has experienced above normal mean temperatures and rainfall amounts, as well as unusually heavy rainfall events, during El Niño (or ENSO positive phase) episodes. The resulting floods have affected thousands of individuals, damaged numerous houses, public buildings and highways, and submerged entire crops, cattle grasslands and livestock farms. Recent examples include the catastrophic events that took place in the Paraguay and Paraná rivers, particularly during the El Niño 1982-83, 1991-92 and 1997-98 years. In the flat Paraguayan savannahs, rainfall extreme events have also caused a proliferation of *Aedes aegypti* mosquito breeding sites, increasing the incidence of classic dengue fever. This has been particularly true in the central, northeastern and eastern portions of this country, where climatic conditions are suitable for the successful development of *Aedes* mosquitoes. During La Niña events (or ENSO negative phase), the country usually experiences the opposite (i.e. rainfall deficits and well below normal ambient temperatures). The concomitant droughts cause a decline in dairy production, an increase in the rate of desertification of arable land, and a rise in the occurrence of grassland and forest fires. They also decrease hydropower generation, increase the pollution of rivers and pools of stagnant water, and limit fluvial transportation, thereby increasing import/export transport fees and diminishing the trade of goods. Moreover, long dry spells affect sunflower, maize, soy, cotton, and wheat production, thus reducing the revenue from these key agricultural activities. All these impacts, although mainly the ones affecting the agriculture sector, prompted the creation of a multidisciplinary group led by the Paraguayan Meteorological and Hydrological Service, and the Risk Management Unit at the Ministry of Agriculture and Livestock Farming. Collaboratively, the group issues an agro-meteorological bulletin, every month or at other intervals depending on specific needs, following the approach presented in the decision-making information system below. Activities include (see Figure 11) the

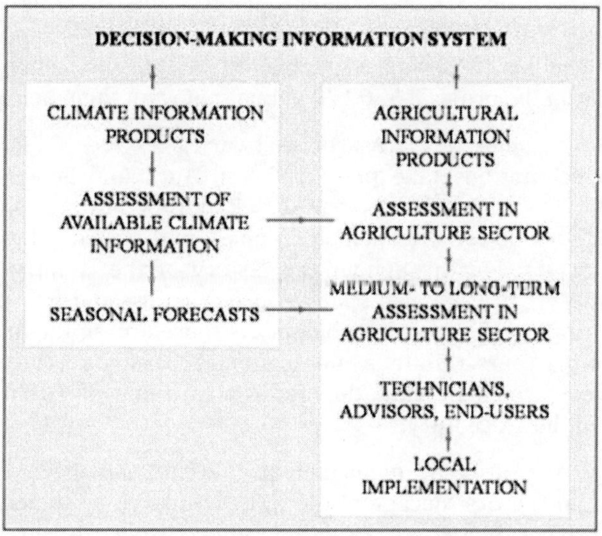

FIGURE 11. The Paraguayan monthly agrometeorological bulletin (available online at: http://www.meteorologia.gov.py/) and the decision-making information system.

assessment of available climate information, seasonal and ENSO forecasts, and medium- to long-term potential impacts on local agricultural production. Key information is then shared with local farmers and smallholders through technicians, advisors and on-the-ground experts.

Up to date, the total number of end-users has reached over 2,000 local, regional and international groups including, among many others, the South American Common Market (MERCOSUR) and organizations in the United States of America, Canada and European countries. Numerous technicians, decision-makers, researchers, and students also use this sectoral information in their individual routine activities.

5.6. Climate Risk of Droughts in Chile And Its Effect on Agriculture

In central Chile, the *secano* zone involves a total surface of 4.362 km^2, and a human population of 54.450 habitants (6.25% of the regional population). Considered as a rural area, the *modus vivendi*depends basically on agriculture. In the coastal secano of the O´Higgins Region, the total surface designated to crops and plantations is on the order of 22.800 hectares. The majority of this land possesses forage plants (27%), cereals –especially wheat- (21%), fruits (20%), vines and vineyards (25%), and in lesser proportion legumes (5%) and vegetables (2%) [26]. Presently, the cattle production is on the order of 536.170 animals, most of them being sheep.

The secano zone shows a dry season that varies between 6 and 8 months per year. Precipitations take place between April and September, with values around the 500-600 mm/year [27]. Reports on droughts in Chile go back to the times before the Spanish Colonies, with a strong impact in the agriculture production throughout the history of the country [28]. In a period of 400 years, 25% has been reported as dry and half of them as extremely dry [29]. The mean probability is therefore on the order of one dry year every 4 years. Quintana and Aceituno [30] have recently discussed trends of decrease in precipitation for central Chile of 10-30% for the second half of the XX century.

Traditionally, the drought management in Chile has been contingent. Nonetheless, in the first decade of the XXI century, government actions in this regard have increased, orienting efforts towards an integrated management of extreme climate events. For example, this originated the National Plan for Civil Protection in 2002. The formal approval of the National Climate Change Strategy in 2006 and the respective Action Plan in 2008 have revealed the adaptation needs of the different productive sectors of Chile, but also the most vulnerable territories and populations. All this created a positive scenario for the establishment of improved policies for climate risk. Three important examples after the extreme drought of 2008 are the creation of the Comité Interministerial de Recursos Hídricos (Interministry Water Resources Committee), the Comisión Nacional de Emergencias Agrícolas y Riesgos Agroclimáticos (National Committee for Agriculture Emergencies and Agro-Climate Risk) and the Sistema Nacional de Gestión del Riesgo Agroclimático (the National System for Agro-Climate Risk Management, found at http://www.minagri.gov.cl/agroclimatico/comision_nacional.php).

The public risk management initiatives implemented by the Chilean organizations involve one or several of the following mechanisms. While excellent initiatives, some of them can still be further improved. Some suggestions are pointed out in those cases.

Early Warning System

Chile possesses a modern early warning system that is able to detect and send alerts about natural disasters. It uses information produced by the Dirección Meteorológica de Chile (DMC), Dirección de Obras Hidráulicas (DOH), Dirección General de Aguas (DGA), Servicio Hidrográfico y Oceanográfico de la Armada (SHOA) and others. The Ministry of Agriculture via its Regional Centers for Agro-meteorological Information interacts with the climate information providers and generates tailored products for the corresponding sector. Nonetheless, these products do not always arrive on time to the farmers.

Agriculture Emergency Announcements

The Agriculture Emergency Zone (ZEA, in Spanish) announcements for droughts are fast and the corresponding institutional instances are efficient. However, the criteria used for announcing the zones, which is based on the agriculture impact, may require modifications. The impacts are assessed in terms of a combination of factors: e.g. losses reported by a technical organism, request of a local authority and help requests from affected farmers. Specifically, ZEA announcements may require a criteria homogenization for both monitoring the evolution of the drought, as well as the emergency declaration itself. The announcement must adjust to regional realities and take into consideration the potable water supply and its effects on agriculture activities.

Budget Availability for Emergencies

Even when the budget allocation for emergencies is normally made through the utilization of funds already directed to other activities, in the last years, there has been an increasing amount of money in Chile properly directed to environmental emergencies. For instance, the emergency drought budget made available for the 2007-2008 case enabled the development of a higher number of implemented policies. In the Ministry of Agriculture, different mechanisms (e.g. emergency bonds, accident benefits, livestock health operations, agriculture emergency employment programs and a special incentive fund program for the recovery of degraded soils) were able to take care of around 161,000 beneficiaries.

A final comment can be included here about the implementation of an additional system to monitor the success of the risk management process in Chile. Presently there is neither an evaluation of the loss levels associated with a drought, nor an *ex post* diagnostic of the implementation mechanisms. It is necessary to possess adequate metrics to assess the effectiveness of applied strategies and the recovery of the affected populations. It is important to develop and to implement a protocol for the evaluation of the impacts of droughts. In addition, this will help with emergency management policies and the development of *ad hoc* prevention and mitigation measures focused on the vulnerability reduction of certain territories. A continuously updated registration of the users and affected people during an emergency is also required; this information will help to project estimates of the offer and demand under similar circumstances. The associated database will consolidate the emergency monitoring system and also will provide useful information for future studies and experience-sharing with other institutions.

6. DISCUSSION AND CONCLUDING REMARKS

The Latin American Observatory partnership has been able to improve several aspects of the way decision-making tools are created and provided to stakeholders and end-mile users in different institution of Latin America in the last years, increasing the dissemination of available services for assessing and –when possible- forecasting hazards and vulnerabilities. This allows advisement of authorities in taking the right course of action in order to protect human populations from the direct or indirect effects of hazards.

In this chapter, we have presented the general methodologies that are used in the OLE^2 for risk assessment and management, along with a few case studies for different countries. The main innovation is related to the way the partnership interacts to share data, products, experiences, helping institutions with lesser human resources and capacities to take advantage of the strengths of other partners, but also sharing efforts at a regional level so the work is divided among a greater number of experts, even when they don't belong to a certain institute, or even when they are not present in the same country. This interdependence has proven a key component of the long-term sustainability of the Observatory.

In terms of the methods employed, important learned lessons arise in relation to the use of probability maps when working with vulnerabilities, hazards and risk management. A probability approach involves the consideration of the related uncertainties of every step of the methodology, which are always present in this kind of analysis. This introduces an *a priori* spectrum of possible immediate actions to decision-makers, even if the real intensity of the hazard is not known until the moment of the emergency. The whole system is aimed at improving the response times and also benefit/cost ratios.

Providing services that take into consideration different scales (e.g. interannual to decadal to climate change scales, as well as from basins to nation-wide to continental scales) and the evolution in time of both hazard and vulnerability probabilities has proven in our experience to be a more adequate practice when assessing and managing risk.

The recognition of the local chains of dissemination of the information, institutional relationships and their adequate use and promotion instead of creating new ones is another learned lesson. The Observatory as a "boundary institution" is paying special attention to these issues and also to the importance of "translating" methods and products so the final services can not only be adequately understood but also be properly assimilated by decision-makers and stakeholders.

Finally, there is an increasingly growing culture on risk assessment and management in LAC, but as Baethgen has pointed out recently [31], it is also important to consider the benefits that may arise if there is also preparedness to take advantage of probable "positive" natural events (e.g. a natural event may be a hazard for certain sectors of production, but may be positive for others). The Observatory must therefore consider also Opportunity Assessment and Management policies in the near future to address such profitable scenarios.

NOTES

[1] - http://ole2.org

[2] - http://mediawiki.cmc.org.ve

[3] - http://laft.cmc.luz.edu.ve:8080/foro_ole2

[4] - ole2@cmc.org.ve

[5] - CRRH is the technical Secretariat of the Central America Integration System, responsible for the coordination of activities related to weather forecast, climate, water resources and climate change assessments with Belize, Costa Rica, El Salvador, Guatemala, Honduras, Nicaragua and Panamá.

REFERENCES

1. Á. G. Muñoz, A. Núñez, R. Cova, System. Climate, An. Simulations, Multi. Integrated-Scale, for. Approach, Research, Decision-Making, Integrated, Multi-Scale Approach for Research and Decision-Making. In Computational Simulations and Applications, Chapter 21. Zhu, J. (Ed.), 978-9-53307-430-6InTech Publishing. 4494682011
2. Ángel. G. Muñoz, Coauthors, An Environmental Watch System for the Andean Countries: El Observatorio Andino. Bull. Amer. Meteor. Soc., 20109116451652
3. Á. G. Muñoz, X. Chourio, S. Reverol, A. Urdaneta, C. Díaz, A. N. D. E. S. G. R. I. D. A. Grid, for. Insfrastructure, in. Geosciences, Andes. the, Proc. Latin-American Conf. on High Performance Computing, Gramado, Brazil, Bull, Hewlett-Packard, Intel Corporation, Microsoft, Silicon Graphics International, 2010914
4. S. Mora, Disasters are not natural: risk management, a tool for development. From: Culshaw, M. G., Reeves, H. J., Jefferson, I. & Spink, T. W. (eds) Engineering Geology for Tomorrow's Cities. Geological Society, London, Engineering Geology Special Publications, 200922101112
5. C. . Vaughan, Climate Service Partnership's Program Manager), personal communication.
6. S. Mora, K. Keipi, Disaster risk management in development projects: models and checklists. Bulletin of Engineering Geology and the Environment, 2006200665155165
7. O. Cardona, Indicators of disaster risk and risk management in Latin America and the Caribbean. Instituto de Estudios Ambientales (IDEA). Department of Sustainable Development, Inter-American Development Bank, Washington, DC. 2006
8. John. L. Beven, Lixion. A. Avila, Eric. S. Blake, Daniel. P. Brown, James. L. Franklin, Richard. D. Knabb, Richard. J. Pasch, Jamie. R.

Rhome, Stacy. R. Stewart, Atlantic Hurricane Season of 2005. Monthly Weather Review, 2008136311091173

9. L. Goddard, W. Baethgen, A. Greene, R. Seager, P. Gonzalez, Á. G. Muñoz, A. Ines, C. Tebaldi, A. Ruane, Multi-scale Climate Information for Agricultural Planning in Southeastern South America for Coming Decades. NSF-EaSM. 2012
10. G. Wieczorek, M. Larsen, L. Eaton, B. Morgan, J. Blair, Debris-flow and flooding hazards associated with the December 1999 storm in coastal Venezuela and strategies for mitigation. U. S. Geological Survey. Open File Report 0101442006
11. L. Guenni, A. Hernandez, M. Fillipone, Modelling population vulnerability and risk to extreme rainfall events in Venezuela. Acta Científica Venezolana, 54Sup. 1: 2-12. 2003
12. W. Torres, Á. G. Muñoz, Simulación Computacional con Modelos Dinámicos del Evento Extremo de Vargas (Venezuela), 1999. BSc Thesis. Universidad del Zulia. 2011pp.
13. W. Torres, Muñoz. Á. G. And, Warning. Early, Risk. Flood, for. Assessment, state. Vargas, Venezuela, In preparation.
14. A. J. Mc Michael, D. H. Campbell-Lendrum, C. F. Corvalan, K. L. Ebi, A. Githeko, J. D. Scheraga, A. Woodward, Climate change and human health: Risks and Responses. World Health Organization, Geneva. 322 2003
15. D. Ruiz, S. Connor, M. Thomson, A. multimodel, in. framework, of. support, surveillance. malaria, control, Chapter VII, 101125in: Seasonal Forecasts, Climatic Change, and Human Health- Health and Climate / Advances in Global Change Research 30Madeleine C. Thomson, Ricardo Garcia Herrera and Martin Beniston (ed.), Springer Science + Business Media, Dordrecht; Publisher: Springer Netherlands, 978-1-40206-876-8The Netherlands. 2008
16. Conservation Internacional.Integrated National Adaptation Pilot project results- Final report (in Spanish). 122 pages. Available online at:http://www.conservation.org.co/.2011
17. Instituto de Hidrología, Meteorología y Estudios Ambientales-IDEAM, Colombia.Segunda Comunicación Nacional ante la Convención Marco de las Naciones Unidas sobre Cambio Climático. Bogotá, Colombia. 2010

18. M. C. Thomson, S. J. Connor, The development of Malaria Early Warning Systems for Africa. Trends in Parasitology 1794384452001
19. G. Poveda, W. Rojas, M. L. Quiñones, I. D. Vélez, R. I. Mantilla, D. Ruiz, J. S. Zuluaga, G. L. Rúa, Coupling between annual and ENSO timescales in the malaria-climate association in Colombia. Environmental Health Perspectives 1094894932001
20. M. C. Thomson, F. J. Doblas-Reyes, S. J. Mason, R. Hagedorn, S. J. Connor, T. Phindela, A. P. Morse, T. N. Palmer, Malaria early warnings based on seasonal climate forecasts from multi-model ensembles. Nature 4395765792006
21. D. Ruiz, G. Poveda, I. D. Vélez, M. L. Quiñones, G. L. Rúa, L. E. Velásquez, J. S. Zuluaga, Modelling entomological-climatic interactions of Plasmodium falciparum malaria transmission in two Colombian endemic-regions: contributions to a National Malaria Early Warning System. Malaria Journal 5:66, doi:10.1186/1475-2875-5-66.2006
22. G. Poveda, M. L. Quiñones, I. D. Vélez, W. Rojas, G. L. Rúa, D. Ruiz, J. S. Zuluaga, L. E. Velásquez, M. D. Zuluaga, O. Hernández, Desarrollo de un Sistema de Alerta Temprana para la malaria en Colombia. Universidad Internacional de Andalucía. 182 2008
23. D. Ruiz, A. M. Molina, M. L. Quiñónes, M. M. Jiménez, M. Thomson, S. Connor, M. E. Gutiérrez, P. A. Zapata, C. López, A. Londoño, Simulating malaria transmission dynamics in the pilot areas of the Colombian Integrated National Adaptation Pilot project. Escuela de Ingeniería de Antioquia, 374 2011
24. Á. G. Muñoz, C. Recalde, Reporte metodológico sobre el experimento de predicibilidad de malaria en el Litoral Ecuatoriano. Proyecto INAMHI-MAE-SCN-PRAA-PACC. 52 2011
25. A. G. Muñoz, D. Ruiz, C. Recalde, Malaria biological models and dynamical downscaling for northwestern South America in the Observatorio Andino framework. ICID+18 / 2nd International Conference: Climate, Sustainability and Development in Semi-arid Regions, Fortaleza, Ceara (Brazil). 2010
26. INE-Instituto Nacional de Estadísticas de Chile2007VII Censo Agropecuario y Forestal 2007.
27. Dirección Meteorológica de Chile (DMC).Climatología Regional. Informe Técnico. 47 2001

28. Urrutia de Hazbun, Rosa y Carlos Lanza Lazcano.Catástrofes en Chile 1541-1992. Santiago, Editorial La Noria. 440 1993
29. Norero, Aldo y Carlos Bonilla (ed).Las sequías en Chile: causas, consecuencias y mitigación. Colección en agricultura, Facultad de Agronomía e Ingeniería Forestal, Universidad Católica de Chile. 128 1999
30. J. Quintana, P. Aceituno, Changes in the rainfall regime along the extratropical west coast of South America (Chile): 3043S. Atmósfera 25 (1), 1-22. 2012
31. W. Baethgen, Climate Risk Management for Adaptation to Climate Variability and Change. Crop Sci. 50:S-70S-76. 2010

7

CHAPTER

VARIABILITY OF THE COURSE OF TOMATO GROWTH AND DEVELOPMENT IN POLAND AS AN EFFECT OF CLIMATE CHANGE

Robert Kalbarczyk[1], Beata Raszka[2] and Eliza Kalbarczyk[1]

[1] Department of Meteorology and ClimatologyWest Pomeranian University of Technology in Szczecin, Poland
[2] Department of Spatial Economy, Wrocław University of Environmental and Life Sciences, Poland

1. INTRODUCTION

The course of phenological phases play an important role in the shaping of yield quantity and quality (Mozny et al., 2009; Peiris et al., 1996; Tao et al., 2006). The length of the development stages is important for the proper formation of both vegetative and reproductive organs. The main meteorological factor affecting the rate of plant development is air temperature (Ahmed et al., 2004;Chmielewski et al., 2005; Popov et al., 2003; Schleip et al., 2009b; Sysoeva et al., 1997).

Since the mid-20th century significant changes in temperature values have been observed in the growing season of crop plants. For instance, in the years 1961-2000, the average increase in air temperature in Germany, in the February-April period, amounted to 0.41°C/10 years (Chmielewski et al., 2004). In Poland, an increase in average air temperature during each April-October period in 1973-2002, on average, amounted to 0.54°C/10 years (Kalbarczyk E. & Kalbarczyk R., 2010). Similarly, positive temperature trends have been confirmed for the growing seasons of, e.g., onions and cucumber (Kalbarczyk, 2009a, 2009b, 2010a). Positive trends of air temperature in the growing seasons of crop plants have also been confirmed in other parts of the world (Bonofiglio et al., 2009;Matsumoto, 2010; Parey, 2008; Peng et al., 2004).

Since the mid-20th century changes in air temperature values have had influence on the course of the growth and development of plants. All over the world research studies focused on the reactions of fruit trees (Chmielewski et al., 2004; Fujisawa & Kobayashi, 2010), wild-growing plants (Gordo & Sanz, 2009; Kalvāne et al., 2009;Moiseev et al., 2010; Yoshie, 2010), and crop plants (Ahas et al., 2002;Dalezios et al., 2002; Kalbarczyk, 2009a; Mazurczyk et al., 2003; Menzel, 2000) have been conducted. Changes in temperature values lead to changes in duration of particular stages and the whole growing season of plants (Peiris et al., 1996; Song et al., 2008; Tao et al., 2006). Shifts in the course of the phenological phases may be radically different. The phenological phases are influenced by climate change, and depend on the species and a region of the world. Differences include acceleration to time delay of the date of a phenophase (Chmielewski et al., 2004;Jorquera-Fontena & Orrego-Verdugo, 2010; Lobell et al., 2007; Wang et al., 2008; Xiao et al., 2008). In the summer, a 1°C increase in the minimum temperature resulted in acceleration of the date of maize flowering by 4.2 days (Tao et al., 2006). In Poland, an increase in the average April temperature by 1°C caused acceleration of the emergence date by 2.5 days (medium-early potato) or 1.7 days (medium-late potato) (Kalbarczyk E. & Kalbarczyk R., 2010). An increase in the average May temperature by 1°C caused acceleration of flowering by about 2.5 days for both cultivars of potato. According toChmielewski et al. (2004) in Germany a 1°C increase in air temperature in the period from February to April caused the beginning of the growing season and flowering of fruit trees to accelerate by about 5 days and the beginning of winter rye shooting to accelerate by 3.8 days. On the other hand, in north-west China, an increase in the minimum temperature by 1°C caused the analysed development stage of cotton to lengthen by 12 days (budding - anthesis) and 9 days (anthesis - boll - opening) (Wang et al., 2008).

Acceleration in the course of plant development caused by a temperature increase, most often leads to reduction in yield quantity. The negative influence of rising temperature, in the period of plant growth and development was discovered, e.g. in the case of winter wheat and rice (Peng et al., 2004; Tao et al., 2006; Wang et al., 2008). Rising temperature, however had a positive effect on the quantity of maize yield in north-east China (Tao et al., 2006). According to the research conducted with the use of simulation models, the influence of rising temperature on crop plant yields may be diverse, as the kind of plant and region of cultivation must be taken into account. Both reduction in yield quantity, as in the case of winter and spring wheat and beans, (Peiris et al., 1996; Wang & Connor, 1996), and its increase, as in the case of potato in the EU, are possible (Peiris et al., 1996; Wolf & Oijen, 2002). Differences in the harvested yield caused by temperature change may be minor or may be a several dozen % yield difference.

The relationship between temperature change and changes in the phenology of crop plants have been researched. The studies on the field cultivation of vegetable plants, however, are relatively rare.Considering the size of production, calculated by means of the yield volume, tomato ranks second to potato, globally. In 2007, over 126.5 m tonnes of tomato were harvested in the world (FAO, 2008). The leading producers among the European Union countries are: Italy (6.0 m tonnes), Spain (3.7 m tonnes) and Greece (1.5 m tonnes). Poland ranks eighth in the EU, with a tomato production of 0.25 m tonnes. Crop plant cultivation plays an important role in Polish vegetal production. In the domestic structure of cultivated vegetables the share of tomato is relatively small, namely about 6% (GUS, 2010). The high thermal requirements of this plant are the reason for its low ranking. The optimum growth temperature is within a range of22-27°C during the day and 16-18°C at night, and is determined by tomato development stage and light intensity. Temperatures above 30°C and below 10-12°C constitute the so-called developmental maximum and minimum. Reduction in yield is already observed when the temperature exceeds 25°C (Tshiala & Olwoch, 2010). Tomato is sensitive to the cold (0-5°C) and frost. When temperatures drop below 0°C, the plants freeze and die. At least a 4 month frost-free duration period is needed (Babik, 2004). Thus, in Poland, the climate thermal requirements are limited.A wide strip of central Poland, the Wrocław region, and the Sandomierz-Lublin region, situated in the south, are considered the most favourable areas for ground cultivation (Skąpski & Borowy, 2000). The best yield is harvested in the years which are considered by Polish terms, as warm and dry. In Poland, tomato yield slightly exceeds 20 $t \cdot ha^{-1}$ (GUS, 2008).

The domestic cultivated surface of the plant since 2001 has remained almost at the same level. Predicted adaptation of Polish agriculture to climate changes till 2030, depending on the direction of the changes, show the possibility of a considerable lengthening of the farming season. The farming season is defined as the period when doing field work is possible (Kundzewicz & Kowalczak, 2008). A 2-3 month lengthening of the climatic growing season of plants, and more of a possibility for thermophilous plant cultivation are also forecast. Therefore, more years with favourable weather conditions for good yield of field-cultivated tomato is possible.

The first goal of the undertaken study was to determine changes occurring in the course of tomato growth and development. The second goal was to determine tomatoes' dependence on air temperature during its growing season.

2. MATERIAL AND METHODS

The results of field experiments on tomato (*Lycopersicon esculentum* Mill), carried out in 22 stations of the Research Centre for Cultivar Testing (COBORU) in Poland in the years 1965-2004 were used in this study. Our study was based on data concerning tomato growth and development. Agrotechnical dates used were: planting up (Pu), the beginning of harvesting (Bh) and the end of harvesting (Eh). Phenological dates used were: the beginning of flowering (Bf) and the beginning of fruit-setting (Bfs). In order to standardise the description of tomato development, the development stages (Bf, Bfs) determined according to the instructions of COBORU were additionally characterised by means of the numerical scale BBCH (Meier,2001). This scale is employed in European Union countries. A method of determining development stages of monocotyledonous and dicotyledonous plants is used. The analysis in the study also included tomato development stages; four short ones: planting up – the beginning of flowering (Pu-Bf), the beginning of flowering – the beginning of fruit-setting (Bf-Bfs), the beginning of fruit-setting – the beginning of harvesting (Bfs-Bh), the beginning of harvesting – the end of harvesting (Bh-Eh); and a long one: planting up – the end of harvesting (Pu-Eh).

The experimental data of COBORU were collected for all the most commonly cultivated varieties of dwarf, flexible-stemmed tomato examined in a given year. After averaging, the data were accepted as a collective standard of the described plant.The use in the research of the collective standard was based on an assumption that intra-species differences do not obfuscate the sought after general regularities of the species. The field experiments took place in the whole area of Poland, except the submountainous regions located in the south-west and south-east of the country. The submountainous regions were excluded from the analysis on the grounds that field cultivation in Poland does not usually occur 500 metres above sea level.

Field experiments in the years 1965-2004 were conducted according to the methodology of COBORU (Domański, 1998). Tomato was cultivated on soils typical for this plant, i.e. soil rich in nutrients, not very heavy and easily warmed up. Depending on present soil richness, mineral fertilization application fluctuated from 150 to 625 kg per 1 hectare of crop. The average mineral fertilization amounted to 405 kg per 1 hectare of the crop, including N and P_2O_5 which were sown at 120 and 95 kg respectively, and K_2O – at 190 kg each. When full autumn organic fertilization with well-decomposed organic manure or compost at a dose from 20 to 30 t ha^{-1} was used for tomato cultivation, the dose of mineral fertilizers was reduced to 200 kg of NPK per 1 hectare.

In the years 1965-2004, average air temperature data in the period from May to October were collected from all meteorological posts operating at the experimental stations of COBORU. If there was no meteorological post at the location of the tomato experiments, the results coming from a meteorological station of the Institute of Meteorology and Water Management (IMGW) were used in the analysis. The selected IMGW station was situated closest to the COBORU station and best reflected the weather conditions of the conducted experiments. In addition, to determine spatial variability of air temperature, the research used data from 52 stations of IMGW, evenly distributed throughout Poland.

Agrotechnical dates, the dates of phenological phases, tomato development stages, and also thermal conditions of air were characterised with the use of the following statistical indexes: multi-annual average, the value of the highest and the lowest average, absolute minimum and maximum values, the range and standard deviation. Multi-annual average and standard deviation in the 40-year research period were calculated on the basis of data from all considered experimental stations of COBORU, or all meteorological stations of IMGW for the dates and duration of tomato development stages and in the case of air temperature. The range was determined between the highest average value (longest, latest) and the lowest average value (shortest, earliest). Temporal and spatial distribution of the change in the course of tomato development and air temperature were also determined on the basis of their linear trend, determined on the basis of linear regression equation. The study calculated deviations from the average, in the subsequent years of the analysed multi-annual period, in relation to the accepted basic period 1965-2004. Identification of the course of tomato development was carried out on the basis of two statistical parameters: the arithmetic mean and standard deviation.The parameters were determined for the basic period 1965-2004. Criteria and classes are presented in Table 1.

TABLE 1. Classification scale of the agrotechnical dates and duration of tomato development stages in the years 1965-2004.

Class	Criterion	Date	Duration of development stages
1	$\xi > \Pi + 2.0\delta$	anomalously late	anomalously long
2	$\Pi + 1.5S < \xi \le \Pi + 2.0S$	very late	very long
3	$\Pi + 1.0S < \xi \le \Pi + 1.5S$	late	long
4	$\Pi - 1.0S \le \xi \le \Pi + 1.0S$	normal	normal
5	$\Pi - 1.5S \le \xi < \Pi - 1.0S$	early	short
6	$\Pi - 2.0S \le \xi \le \Pi - 1.5S$	very early	very short
7	$\xi < \Pi - 2.0S$	anomalously early	anomalously short

For example, the date of the beginning of tomato harvesting was considered average (normal) when the date in a given year fulfilled the following condition: $\Pi - 1.0S \le \xi \le \Pi + 1.0S$, where Π denotes the average date, δ the value of standard deviation, with both parameters calculated for the basic period 1965-2004.ξ denotes the date recorded in a given year. Similar classifications, but concerning thermal and precipitation conditions were made, among others, by: Pokladníková et al. (2008), Węgrzyn (2007) and Żmudzka (2004).

The relationship between agrotechnical dates and the dates of tomato phenophases, and the average air temperature was determined by means of the simple linear regression analysis. Statistical assessment of the equations was done on the basis of the *t*-Student and *F*-Snedocor tests, the Pearson's correlation coefficient, and a coefficient describing the difference between standard deviation of a dependent variable and a standard error of equation estimation (*S-Sy*) (Dobosz, 2001; Sobczyk, 1998). The occurrence of autocorrelation of random components was checked by means of the Durbin-Watson test. To verify regression equations, the study also used relative forecast error, determined according to the formula:

$$RFE = \frac{y - y_p}{y} \cdot 100\% \tag{1}$$

and average relative forecast error, for all the analysed stations of COBORU and the considered years 1965-2004. An average relative forecast error was calculated according to the formula:

$$ARFE = \frac{1}{n}\sum_{i=1}^{n}|RFE| \qquad (2)$$

where:

y – actual date (day),

$y\,p$ – date calculated according to the formula (day),

n – number of years accepted in a time series (number of stations x number of years).

Also determined, were how many times relative forecast error in the analysed multi-annual period 1965-2004 amounted to |RFE|≤2%|RFE|≤2% and 2%|RFE|≤4%2%|RFE|≤4%.

3. RESULTS

3. 1. Temporal and Spatial Variability of Tomato Growth and Development

3.1.1. Agrotechnical Dates and Phenological Phases

In Poland, in the years 1965-2004, the average date of planting up (Pu) tomato seedlings in the field fell on 21st May (Table 2). The absolute minimum date was 3rd May and the maximum 2nd June.The beginning of flowering (Bf) of the described plant was observed on average on 7th June, and the beginning of fruit-setting (Bfs) on 19th June. The average date for the beginning of harvesting (Bh) fell on 21st July, and the end of harvesting (Eh) on 27th September. The earliest average date was 8th July for the beginning of harvesting and 19th September for the end of harvesting. The latest average date – for the beginning of harvesting was 8th August and 6th October for the end of harvesting. Both absolute minimum and absolute maximum dates of the subsequent agrotechnical dates and the dates of tomato phenophases, were generally different by 2-5 weeks from the average dates. For minimum dates, the bigger differences were in the dates: Pu, Bf and Bfs and in the case of maximum dates: Bh and Eh. The standard deviation of agrotechnical dates and the dates of tomato phenophases fluctuated from about 2 to 8 days. The date of tomato planting up was marked by the lowest standard deviation and the date of the beginning of harvesting by the highest one.The range of the described dates calculated between the average latest and the average earliest date, like standard deviation, was the smallest for planting up and the biggest for the beginning of harvesting.The range swung from a low 7 to as high as 31 days.

TABLE 2. Statistical characteristics of the agrotechnical[1] dates and the phenological phases of tomato in Poland, in the years 1965-2004.

Agrophase	BBCH scale	Date (day)						Absolute date (day)	
		mean	latest	earliest	range[2]	S	trend (day/10a)	max	min
Pu[1]	-	21-05	25-05	18-05	7	2.0	-0.6*	2-06	3-05
Bf	61601	7-06	15-06	31-05	15	3.6	-0.7*	23-06	17-05
Bfs	71701	19-06	30-06	10-06	20	4.8	-1.1*	9-07	26-05
Bh[1]	-	21-07	8-08	8-07	31	7.8	-3.5***	26-08	28-06
Eh[1]	-	27-09	6-10	19-09	17	4.0	2.1***	27-10	30-08

Pu – planting up, Bf – beginning of flowering, Bfs – beginning of fruit-setting, Bh – beginning of harvesting, Eh – end of harvesting, S – standard deviation, max – absolute maximum date, min – absolute minimum date, * significant at $p \le 0.1$, *** significant at $p \le 0.01$, [2]between the latest and the earliest date.

The analysis of the linear trend of agrotechnical dates and the dates of tomato phenological phases showed a statistically significant, negative temporal tendency. This was a tendency for a year by year acceleration of almost all the considered dates, except for the end of harvesting (Table 2, Fig. 1). The biggest acceleration was found for the beginning of harvesting (-3.5 days /10 years, $p \le 0.01$), and next for the beginning of fruit-setting (-1.1 days /10 years, $p \le 0.1$), definitely the smallest for planting up(-0.6 days /10 years, $p \le 0.1$) and the beginning of flowering (-0.7 days /10 years, $p \le 0.1$). The end of tomato harvesting showed significant delay (2.1 days /10 years, $p \le 0.01$), year by year, in the years 1965-2004. Out of all the analysed tomato dates, the highest correlation coefficient for the linear trend in the years 1965-2004 for all of Poland was determined for the end of harvesting ($r = 0.601$), and next for the beginning of harvesting ($r = -0.52$).

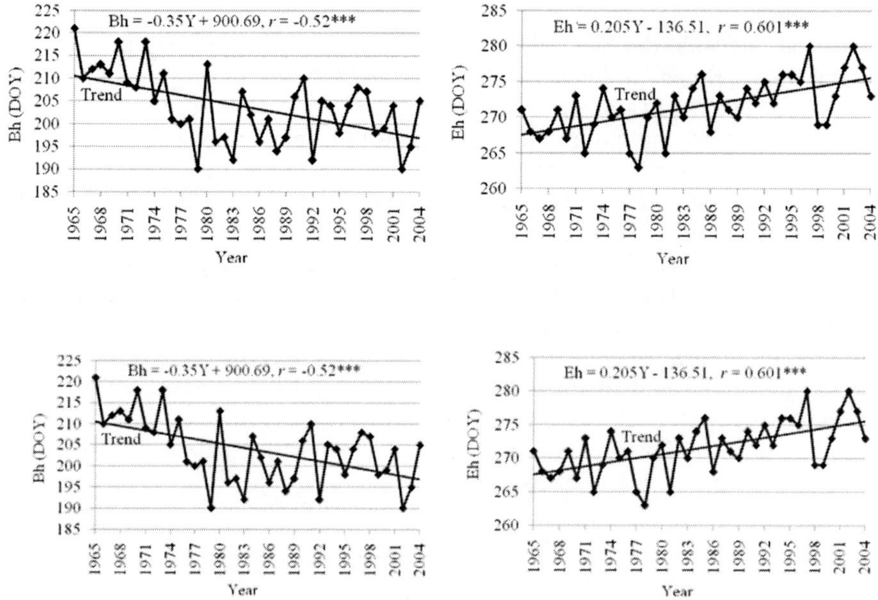

Y - year, r - correlation coefficient, *** significant at $p \leq 0.01$.

FIGURE 1. Temporal distribution of the dates of the beginning of harvesting (Bh) and the end of harvesting (Eh) of tomato in Poland, in the subsequent years of the analysed multi-annual period 1965-2004.

Acceleration of the date of the beginning of tomato harvesting did not occur evenly throughout the whole country (Fig. 2). Significant acceleration of the beginning of harvesting was proved only in the north-west and in the south (over -1.0 day /10 years). The biggest significant acceleration of the beginning of harvesting was in the south-west and in the Kraków region (over -1.5 days /10 years). Delay of the end of harvesting date, like acceleration of the beginning of harvesting, did not have an even occurrence throughout the whole country. Significant delay of the last date of tomato harvesting was recorded in the western and southern parts of Poland, where it usually oscillated from 0.6 day /10 years to 1.0 day /10 years. The biggest delay of the date (over 1.0 day /10 years) occurred in the Kraków and Wrocław regions.

During the 40-year research period, the date of the beginning of tomato harvesting was decidedly delayed 6 times in comparison with the average domestic date, and accelerated 6 times (Fig. 1, Fig. 3). The results are from agrotechnical dates showing that only once in the first half of the research period, in 1965, the beginning of harvesting was anomalously late. It was very late 2 times; in the years 1970 and 1973. It was late 3 times; in 1967, 1968, 1980. The beginning of harvesting was very early in 1979 and early in 1983.In the second half of the research period, delay of the harvesting date was not recorded, but 4 times the date was accelerated. In 1992 and 2002, the beginning of harvesting date occurred very early. The beginning of harvesting date was early in 1988 and 2003. The end of harvesting date of tomato was marked by the opposite temporal structure. In the first half of the analysed multi-annual period, only the years in which acceleration of the date was recorded, were identified, and in the second half of the analysed multi-annual period only the delay of the date years were identified. The end of tomato fruit harvesting was anomalously early in 1978, very early in 1972, 1977, 1981, and early in 1967 and 1970. Tomato harvesting occurred late in: 1985, 1994, 1995, 2001, 2003, and anomalously late in 1997 and 2002.

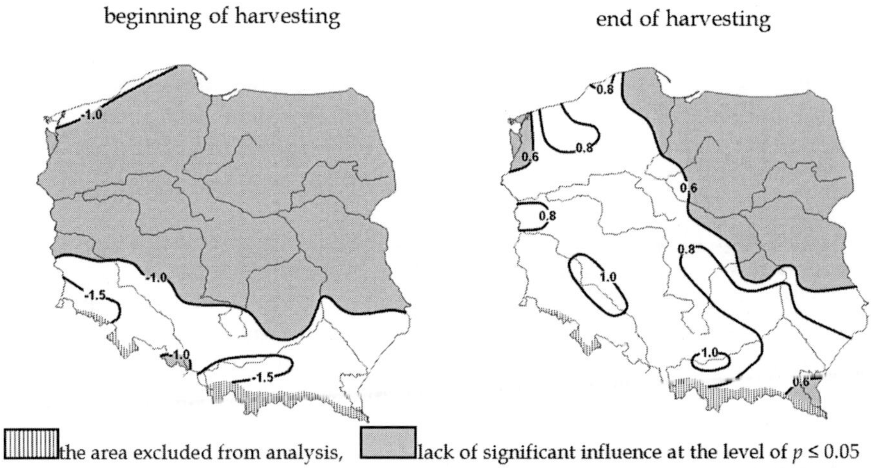

FIGURE 2. Statistically significant, at least at the level of $p \le 0.05$, linear regression coefficients for the trend of the beginning of harvesting date and the end of harvesting date of tomato in Poland, calculated for 10 years, in the years 1965-2004.

Year	Date		Year	Date	
	Bh	Eh		Bh	Eh
1965			1985		
1966			1986		
1967			1987		
1968			1988		
1969			1989		
1970			1990		
1971			1991		
1972			1992		
1973			1993		
1974			1994		
1975			1995		
1976			1996		
1977			1997		
1978			1998		
1979			1999		
1980			2000		
1981			2001		
1982			2002		
1983			2003		
1984			2004		

date:
anomalously late
very late
late
normal
early
very early
anomalously early

FIGURE 3. Identification of the beginning of harvesting date (Bh) and the end of harvesting date(Eh) of tomato in Poland, in the subsequent years of the analysed multi-annual period 1965-2004.

Tomato agrotechnical dates in anomalously extreme years were different from the average domestic date and varied across Poland (Fig. 3, Fig. 4). In 1965, when the date of the beginning of harvesting was identified as anomalously late, deviation from the average multi-annual (1965-2004) date oscillated from less than 4 days to even more than 8 days. The first tomato fruits were harvested in the north and in the south-west of the country at the latest date, 8 days later than the norm. The first tomato fruits were harvested at a slightly earlier date, 4 days later than usual, in the central west, the centre and the south-east. The end of harvesting, especially in 1978, 1997 and 2002 was significantly different from the average domestic date. In 1997 and 2002, deviation from the average multi-annual date for the end of harvesting in Poland, fluctuated mostly from 2 to 6 days in 1997 and from 4 to 8 days in 2002.The biggest delay was recorded in the northern part of the country (> 6 days) in 1997 and in the central-western part (> 8 days) in 2002. In 1978, the last field-cultivated tomato fruits were harvested about 6 to 8 days earlier than the average domestic date. The earliest end of harvesting was recorded in the north-east, north, south-west and locally in the Kielce region.

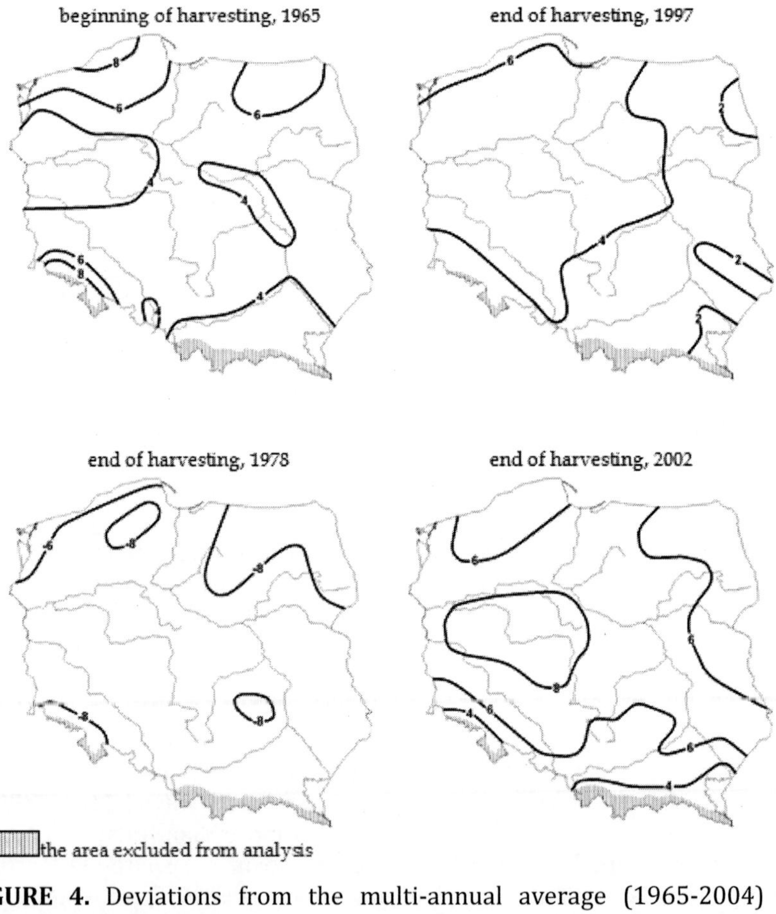

FIGURE 4. Deviations from the multi-annual average (1965-2004) in Poland, of tomato agrotechnical dates: the beginning of harvesting in 1965 and the end of harvesting in: 1978, 1997 and 2002.

3.1.2. Development Stages

On average, the shortest period of tomato development was the period from the beginning of flowering to the beginning of fruit-setting (Bf-Bfs), lasting only 12 days (Table 3). The longest period of tomato development was the period of tomato fruiting, i.e. from the beginning to the end of harvesting (Bh-Eh), which lasted 68 days. Tomato flowering occurs on average, 17 days after the date of planting up in the field. Tomato flowering occurs earliest after 9 days of planting up in the field and latest after 28 days. The beginning of fruit-setting occurred on average 29 days after the date of planting up. In Poland, the beginning of fruit-setting occurred on average, in the last ten days of May (Table 2). The beginning of tomato harvesting occurred on average, 62 days after the date of planting up. The end of harvesting occurred 130 days after the date of planting up. In the analysed multi-annual period there were also years in which the period from planting up to the end of harvesting (Pu-Eh) lasted 139 days, and at particular stations of COBORU it oscillated from 98 to even 161 days. Absolute minimum duration of development stages oscillated from 4 days in the case of the period Bf-Bfs, to 28 days in the case of the period Bh-Eh.

Out of the four considered short periods of tomato development stages, least diverse in terms of duration was the period from the beginning of flowering to the beginning of fruit-setting (Table 3). It was indicated not only by the lowest value of standard deviation (S = 2.0 days), but also by the lowest average value (12 days) and the smallest range (R = 8 days). On the other hand, the periods from planting up to the beginning of flowering (S = 3.6 days) and from the beginning of fruit-setting to the beginning of harvesting (S = 5.2 days) were marked by average diversity of duration. The highest standard deviation (S = 9.5 days) was characteristic of the period from the beginning of harvesting to the end of harvesting. The range between extreme (longest and shortest) average duration periods of tomato development stages, varied from 8 days in the case of the period from the beginning of flowering to the beginning of fruit-setting, to 41 days in the case of the period from the beginning of harvesting to the end of harvesting. In the period from planting up to the end of harvesting, this range amounted to 20 days.

In the years 1965-2004, statistically significant lengthening, at the level of $p \leq 0.01$, of only one out of the analysed four short tomato development stages was seen (Table 3, Fig. 5). In Poland, the period from the beginning of harvesting to the end of harvesting, on average, lengthened by as much as 5.6 days /10 years. In the periods Pu-Bf and Bf-Bfs, no statistically significant examples of shortening or lengthening were found, as opposed to the period Bfs-Bh, which, year by year, became significantly shorter, on average by -1.4 days /10 years. Changes in duration of particular development stages of tomato meant that the whole period lasting from planting up to the end of harvesting, lengthened by as much as 2.6 days /10 years. The correlation coefficient, which was determined for the function best describing the linear trend of development stages, amounted to 0.69 for the period Bh-Eh and 0.65 for the period Pu-Eh.

TABLE 3. Statistical characteristics of duration of tomato development stages in Poland, in the years 1965-2004.

Development stage	Duration (day)						Absolute duration (day)	
	mean	longest	shortest	range[3]	S	trend (day/10a)	max	min
Pu-Bf	17	28	9	19	3.6	n.s.	33	5
Bf-Bfs	12	16	8	8	2.0	n.s.	27	4
Bfs-Bh	33	44	23	21	5.2	-1.4*	80	14
Bh-Eh	68	90	49	41	9.5	5.6***	100	28
Pu-Eh	129	139	119	20	4.7	2.6***	161	98

Pu-Bf - planting up - beginning of flowering, Bf-Bfs - beginning of flowering - beginning of fruit-setting, Bfs-Bh - beginning of fruit-setting - beginning of harvesting, Bh-Eh - beginning of harvesting - end of harvesting, Pu-Eh - planting up - end of harvesting, max - absolute maximum duration, min - absolute minimum duration, n.s. - non-significant at $p \leq 0.1$, [3]between the latest and the shortest duration. The remaining explanations see Table 2.

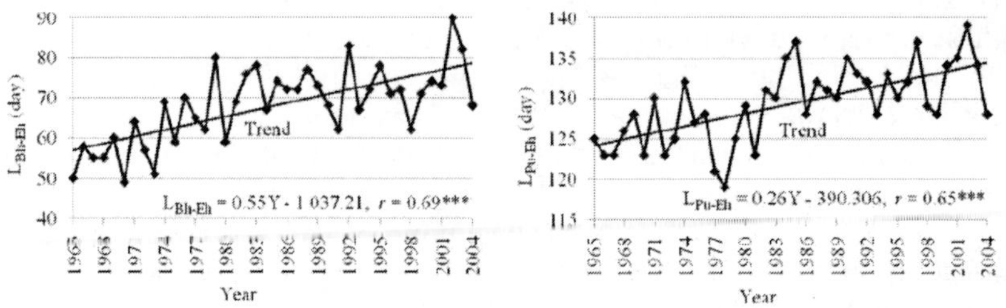

Y - year, r - correlation coefficient, *** significant at $p \leq 0.01$.

FIGURE 5. Temporal distribution of duration from the beginning to the end of harvesting (L_{Bh-Eh}) and from planting up to the end of harvesting (L_{Pu-Eh}) of tomato in Poland, in the subsequent years of the analysed multi-annual period 1965-2004.

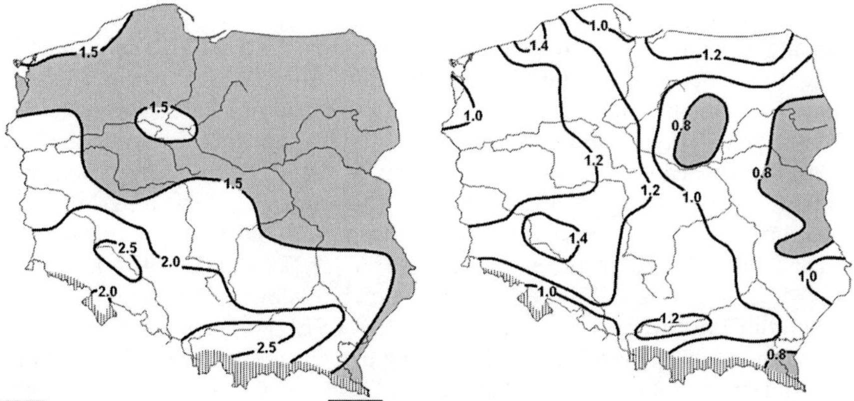

the area excluded from analysis, lack of significant influence at the level of $p \leq 0.05$

FIGURE 6. Statistically significant, at least at the level of $p \leq 0.05$, linear regression coefficients for the trend of duration from the beginning to the end of harvesting and from planting up to the end of harvesting of tomato in Poland, calculated for 10 years, in the years 1965-2004.

Despite the fact that in the whole country, the period from the beginning of harvesting to the end of harvesting of tomato on average, lengthened by as much as 5.6 days /10 years, in particular regions of Poland, this lengthening predominantly oscillated from 1.5 days /10 years to 2.5 days /10 years (Fig. 6). The biggest lengthening occurred in the south of the country, especially in the Wrocław and Kraków regions (Fig. 2). These were the regions where the biggest acceleration of the beginning of harvesting and at the same time the biggest delay of the end of harvesting were recorded. Changes in duration of the period from planting up to the end of harvesting, in particular regions of the country, were also different than for the whole country (Fig. 6). In Poland, lengthening of the period Pu-Eh usually oscillated from 0.8 day /10 years to 1.4 days /10 years. The biggest changes in duration, over 1.4 days /10 years, of the period Pu-Eh, occurred in south-west Poland in the Wrocław region, and in the north-west in the Koszalin region. No proved significant changes in duration of the Bh-Eh periods took place in the northern, north-eastern and eastern parts of the country. No proved significant changes in duration of the Pu-Eh periods took place in the central-eastern part of the country. Spatial distribution of the duration-change of these two tomato development stages was partially consistent with the distribution of the change of these agrotechnical dates: acceleration of the date of Bh and delay of the date of Eh (Fig. 2).

Changes in duration of the periods Bh-Eh and Pu-Eh were also confirmed by identification of these periods, in the subsequent years of the analysed multi-annual period (Fig. 7). Out of the 40 examined years, there were seven years: 1965-1968, 1970, 1972-1973 with shorter than average Bh-Eh periods recorded. In three of the seven years, the Bh-Eh period was identified as very short, and in four years as short. The Pu-Eh period had a similar distribution of years as the Bh-Eh period, but periods shorter than the average were recorded in: 1966-1967, 1970, 1972, 1977-1978 and 1981. In 1978 the Pu-Eh period was even anomalously short and in 1977 – very short.In the remaining five years, the Pu-Eh period was short. On the other hand, in the years 1983-2004, long periods were recorded. The Bh-Eh period in the following six years: 1979, 1983, 1992, 1995, 2002-2003 and the Pu-Eh period in the following seven years: 1984-1985, 1990, 1997, 2001-2003 were longer than the average. Anomalously long Bh-Eh and Pu-Eh periods were recorded in 2002, and very long ones in 1992 and 1985 and 1997.

In Poland, deviation from the multi-annual average (1965-2004) of the length of the period from the beginning to the end of harvestingin 2002, usually amounted to 8 to 14 days (Fig. 8). The period lasted longest, over 14 days, in the central-western part of the country and in the Warsaw region. The period lasted much shorter – less than 8 days, in northern, south-western and south-eastern Poland.In the case of the long development stage of tomato, i.e. Pu-Eh, deviations from the average in 2002 were lower than in the case of the Bh-Eh period. The long development stage of tomato oscillated generally from 4 to 10 days. The Pu-Eh period lasted the longest; over 10 days, in the central-western part of the country. The Pu-Eh period was slightly shorter, less than 6 days, in the north and north-east. The Pu-Eh period was shortest, less than 4 days, in the south-west and south-east of Poland. In 1978, the Pu-Eh period was also different in its duration from the multi-annual average. The Pu-Eh period was the shortest out of the 40 considered periods in the years 1965-2004. In 1978 the Pu-Eh period was the shortest, less than 8 days, in the north, north-east, south-west and locally in the vicinity of Kielce and Mława.

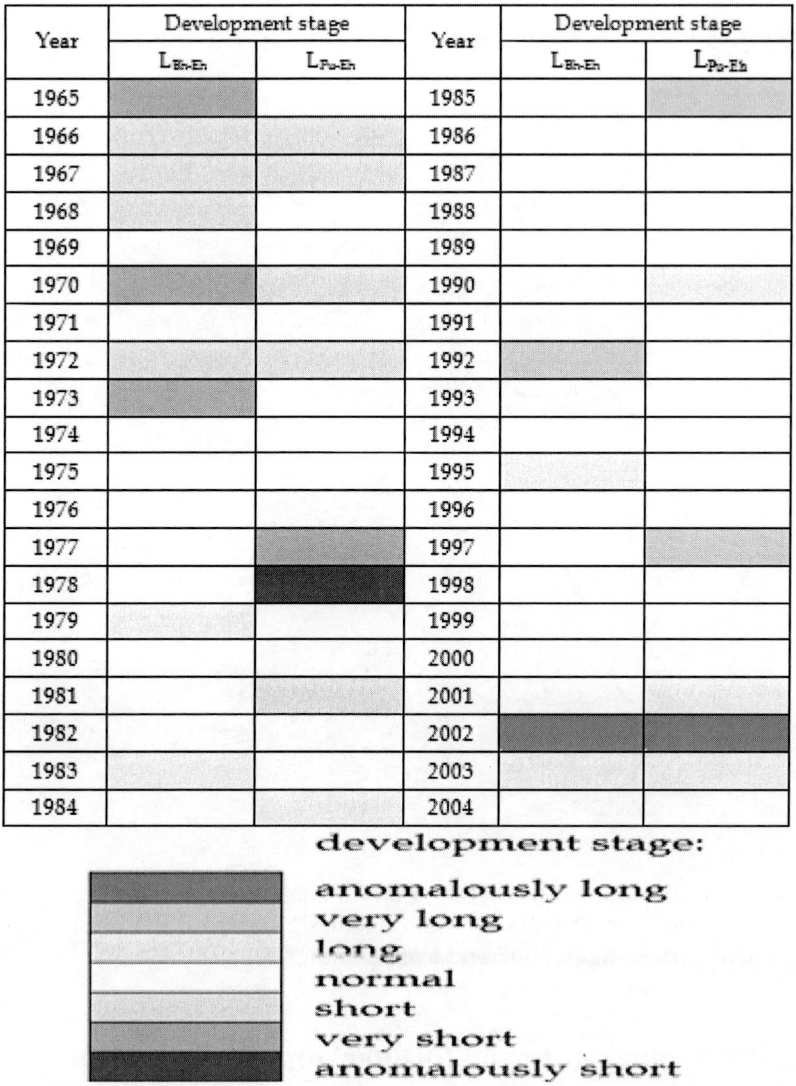

FIGURE 7. Identification of duration from the beginning to the end of harvesting (L_{Bh-Eh}) and from planting up to the end of harvesting (L_{Pu-Eh}) of tomato in Poland, in the subsequent years of the analysed multi-annual period 1965-2004.

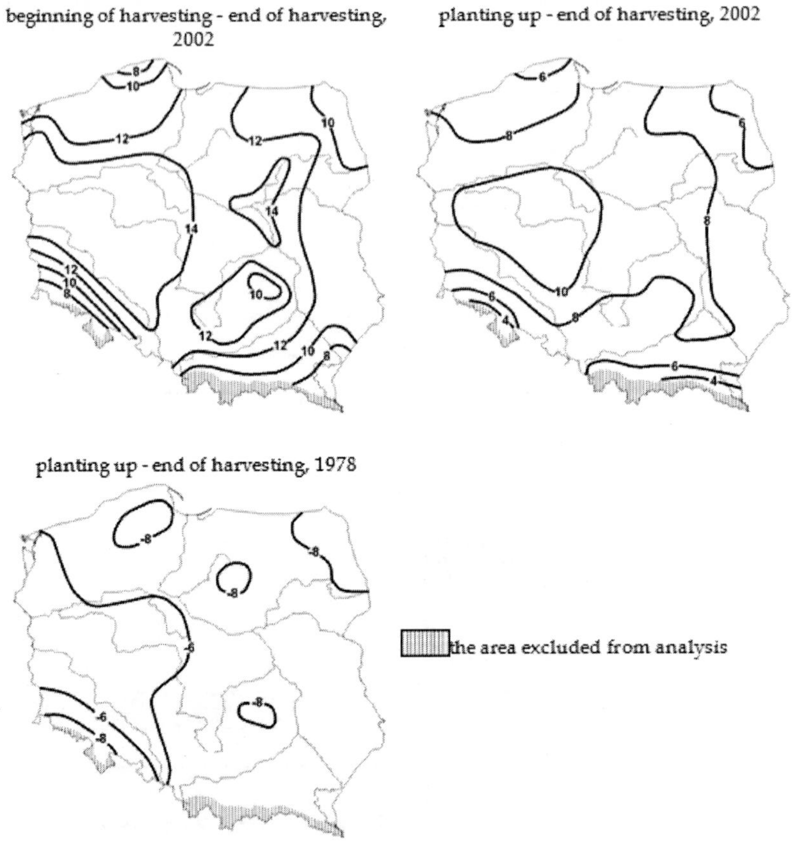

FIGURE 8. Deviations from the multi-annual average (1965-2004) in Poland, of duration of tomato development stages: from the beginning to the end of harvesting in 2002 and from planting up to the end of harvesting in: 1978 and 2002.

3.2. Air Temperature and Its Effect on The Course of Tomato Growth and Development

Changes in the occurrence of agrotechnical dates and the dates of phenological phases of tomato, which occurred during the 40-year research period (Table 2), are closely related to the course of air temperature (Table 4, Fig. 9). Out of the four dates: Bf, Bfs, Bh and Eh, the strongest relationship with air temperature was proved in the case of the phenological phases: the beginning of flowering (r = -0.77, $p \leq 0.01$) and the beginning of fruit-setting (r = -0.82, $p \leq 0.01$). A slightly weaker relationship with air temperature, but also significant at the level of $p \leq 0.01$, was proved in the case of the agrotechnical dates: the beginning of harvesting (r = -0.56***) and the end of harvesting (r = 0.61***). However, apart from thermal conditions of air, agrotechnical dates of vegetables to a large extent, depend on the course of other agrometeorological conditions. Agrotechnical dates of vegetables also depend on other factors, e.g. organisational work on the horticultural farm (Babik, 2004; Domański, 1998).

TABLE 4. Regression analysis describing the relationship between the agrotechnical date (Bh, Eh) and the date of the phenological phase (Bf, Bfs) of tomato and average air temperature in Poland in the years 1965-2004.

Regression equations	Characteristics						
	t	*F*	*r*	*S-Sy* (day)	*ARFE* (%)	frequency of the occurrence of | *RFE* | in range	
	bx / *a*					0-2 (%)	2-4 (%)
Bf = -1.81Ta_V + 182.76	-7.5 / 57.7	56.3	-0.77***	1.4	1.4	77.5	22.5
Bfs = -2.21$Ta_{10V-10VI}$ + 202.72	-8.4 / 55.3	75.3	-0.82***	2.1	1.2	82.5	17.5
Bh = -3.061$Ta_{20V-20VI}$ + 249.62	-4.2 / 22.5	17.6	-0.56***	0.8	2.6	40.0	37.5
Eh = 2.13$Ta_{10VIII-10IX}$ + 237.0503	4.8 / 32.6	22.8	0.61***	1.3	1.8	75.0	25.0

t – value of the *t*-Student test, *F* - value of the *F*-Snedecor test, *r* – correlation coefficient, *S-Sy* - difference between a standard deviation of a dependent variable and a standard error of equation estimation (day), *ARFE* - average relative forecast error (%), *RFE* - relative forecast error (%), *bx* – regression coefficient, *a* – intercept, Ta - average air temperature (°C). The remaining explanations see Table 2.

The best fitted regression function, in relation to empirical data, was the date of the beginning of fruit setting (Table 4). This date is confirmed not only by the highest correlation coefficient but also by the values of the *t*-Student and *F*-Snedecor tests and also the coefficient describing the difference between the standard deviation of a dependent variable and equation standard error (*S-Sy*). The date of the beginning of fruit-setting in Poland can be predicted with average relative forecast error amounting only to 1.2%. The occurrence of relative forecast error in the analysed multi-annual period, 1965-2004 within a range of 0-2%, amounted to as much as 82,5%, and within a range of 2-4% - only 17.5%. Out of the four analysed dates of the described plant, three: Bf, Bfs, Bh are negatively affected by thermal conditions of air, i.e. with an increase in temperature there occurred acceleration of the date. If air temperature in May increases by 1°C, then the date of the beginning of flowering will be accelerated by 1.8

days.When air temperature increases by 1°C in the period from 10th May to 10thJune, then the date of the beginning of fruit-setting will accelerate by 2.2 days. If air temperature increases in the period from 20th May to 20th June, then the date of the beginning of harvesting may even accelerate by 3.1 days. Only the date of the end of harvesting was positively correlated with air temperature in the period from 10th August to 10th September, a drop by 1°C may delay it by 2.1 days.

In 2002, the highest air temperature in May (Ta_V) was recorded, 3.3°C higher than the average (Fig. 9). In 2002, the date of the beginning of tomato flowering was earlier than the average in the years 1965-2004, by about 5 days. In the period from 10th May to 10th June ($Ta_{10V\text{-}10VI}$), air temperature was highest in 1979 (3.7°C higher than the average). Also in 1979, the air temperature was the highest (3.5°C higher than the average) in the period from 20th May to 20th June ($Ta_{20V\text{-}20VI}$). In the period 10th August to 10th September ($Ta_{10VIII\text{-}10IX}$), air temperature was highest in 2002 (3.2°C higher than the average), i.e. like in May. Air temperature in 1979 was higher than the average 1965-2004 air temperature. The higher 1979 air temperature contributed to acceleration of the dates of the beginning of fruit-setting and the beginning of harvesting, respectively by about 9 and 14 days compared to the average date. On the other hand, above-average temperatures in 2002 contributed to about an 8 days delay in the end of harvesting date.

In May, in extreme years in terms of the course of thermal conditions of air, spatial distribution of air temperature was considerably different from the multi-annual structure in the years 1965-2004 (Fig. 10). In 1980 deviations from the norm of the analysed meteorological element, varied in most regions of Poland from -3.0 to -5.0°C, and in 2002 from 1.0 to 4.0°C. In 1980, it was coldest in the north-eastern part of the country and warmest in the central-west. In 2002, the lowest, positive air temperature which deviated from the average, were recorded in the north-west. The highest air temperatures which deviated from the average, were recorded mostly in central and southern Poland. In the periods from 10th May to 10th June in 1991 and 1979, the structures of air temperature deviating from the norm in Poland, were different than in May. In 1991 negative deviations oscillated from above -3.0°C in the central-western and central-eastern parts of the country to below -4.0°C in the northern, south-western and south-eastern parts. In 1979, as opposed to 1991, the highest deviations from the norm, amounting to above 4.0°C, were recorded in the central strip of Poland, stretching in latitude to the eastern border of the country.In the period from 20th May to 20th June, in the years of the highest deviations from the average, air temperature deviation oscillated from below -3.0°C in 1974 to above 4.0°C in 1979 (Fig. 9). In summer, in the period from 10th August to

10thSeptember, the values of temperature deviations from the norm were similar to the earlier characterised periods and they oscillated generally from -3.0°C in 1978 to 4.0°C in 2002. In the summer of 1978, the air temperature was 3.0°C lower than the multi-annual average. This 3.0°C lower temperature occurred mainly in north and south-east Poland (Fig. 10). In the warmest period from 10th August to 10th September, which was recorded in 2002, in the central part of the country, on the Bay of Gdańsk and in the central-western part of Poland. This high 2002 temperature was 4.0°C higher than the average 1965-2004 temperature.In the tomato growing season, in the period from May to October, air temperature in the analysed years oscillated from 12.9 to 15.7°C, and at the stations of COBORU – as much as from 11.6 to 18.0°C. On the other hand, average air temperature during short periods, significantly influencing agrotechnical dates and the dates of phenological phases of tomato, oscillated from 13.0°C for Ta_V to 16.1°C for $Ta_{10VIII-10IX}$(Table 5, Fig. 11). In May, the highest average air temperature amounted to 16.3°C, and was 1.7°C lower than the highest average $Ta_{10V-10VI}$, 2.2°C and 3.0°C lower, respectively, than the highest averages $Ta_{20V-20VI}$ and $Ta_{10VIII-10IX}$. The absolute maximum air temperature was from 8.0°C higher than the minimum one in the period from 10thAugust to 10th September, to 11.2°C in May. Average air temperature in Poland was marked by the highest variability in the period from 10th May to 10th June, which is confirmed by the value of standard deviation (S = 1.8°C). Temperature in the period from 10th August to 10th September was marked by the decidedly lowest variability. Temperature in the period from 10th August to 10thSeptember had the lowest determined standard deviation (S = 1.1°C) and the smallest range (R = 5.8°C).

In the whole growing season of tomato, a significant increase in air temperature by 0.3°C /10 years ($p \leq 0.05$) was proved. Out of the four short analysed periods, there was a significant increase only in two air temperatures, most in May by 0.4°C/10 years ($p \leq 0.01$), and slightly less in the period from 10thAugust to 10th September by 0.3°C /10 years ($p \leq 0.01$). In Poland, air temperature increase was not uniform in all climatic regions (Fig. 12). The changes of agrotechnical dates and tomato development stages were also not uniform in all climatic regions (Fig. 2, Fig. 6). In May, a significant increase in air temperature, by more than 0.5°C /10 years, was recorded in the western part of Poland. The highest increase in air temperature was in the Kraków and Wrocław regions where air temperature increased by as much as 0.6°C /10 years.In summer, in the period from 10th August to 10th September a significant increase of more than 0.4°C /10 years, occurred in the south-west and locally in the central-west and the north. The highest increase of 0.5°C /10 years was in the Wrocław region.

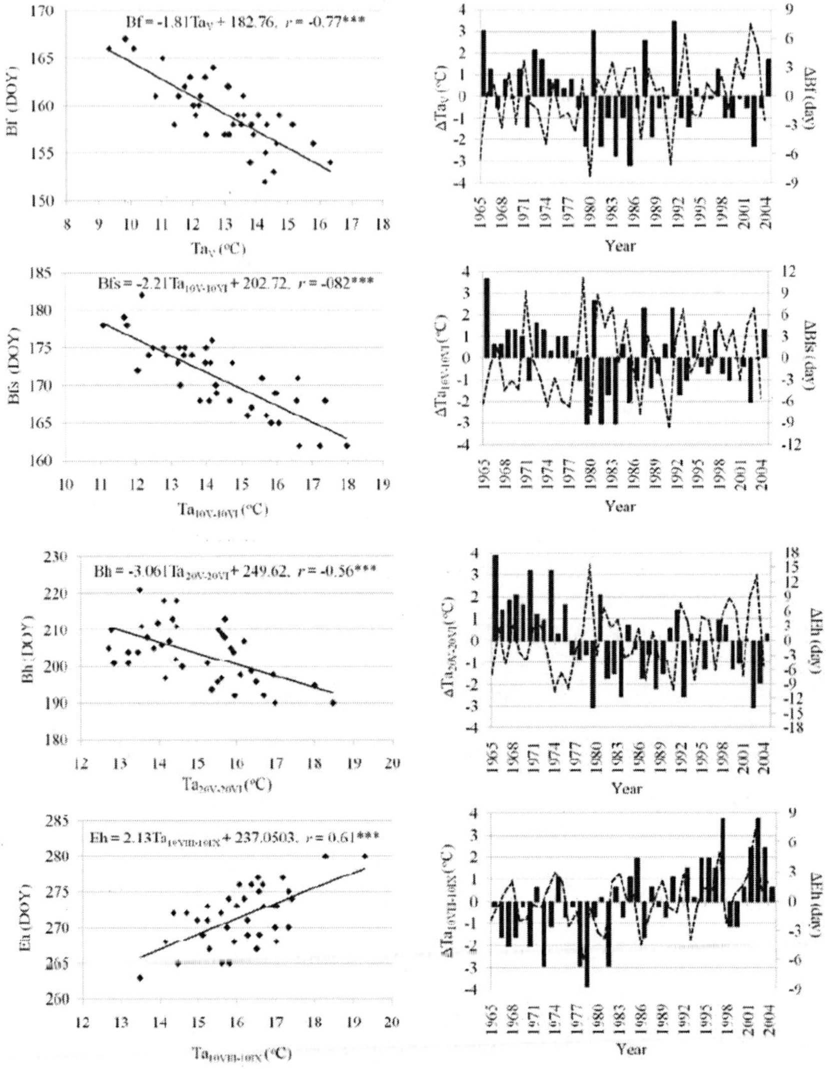

r – correlation coefficient, *** significant at $p \leq 0.01$.

FIGURE 9. Left-hand side: relationship between agrotechnical dates (Bh, Eh)and phenological phases (Bf, Bfs) of tomato in Poland and average air temperature (Tav, $Ta_{10V-10VI}$, $Ta_{20V-20VI}$, $Ta_{10VIII-10IX}$). Right-hand side: deviations from the multi-annual average (1965-2004) in Poland of the dates (ΔBh, ΔEh)and phases (ΔBf, ΔBfs) of tomato and air temperature (ΔTav, $\Delta Ta_{10V-10VI}$, $\Delta Ta_{20V-20VI}$, $\Delta Ta_{10VIII-10IX}$) in the subsequent years of the analysed multi-annual period 1965-2004.

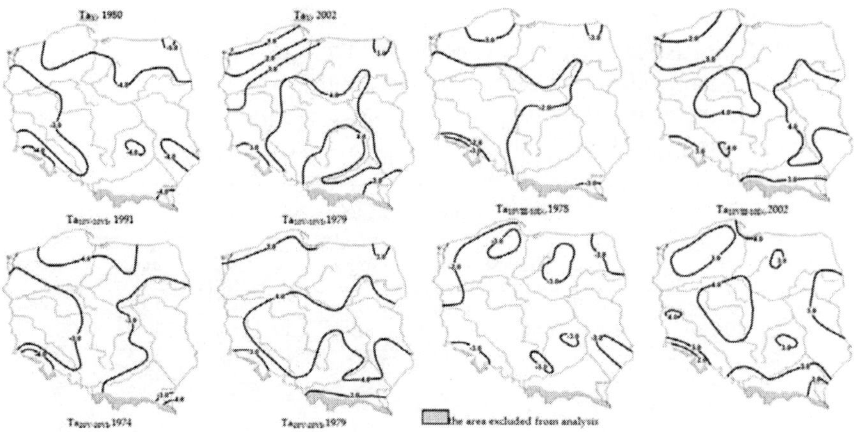

FIGURE 10. Deviations from the multi-annual average (1965-2004) of air temperature in Poland: in May (Ta_V) – in 1980 and 2002, in the period 10th May-10th June ($Ta_{10V-10VI}$) – in 1979 and 1991, in the period 20th May-20th June ($Ta_{20V-20VI}$) – in 1974 and 1979 and in the period 10th August-10th September ($Ta_{10VIII-10IX}$) – in 1978 and 2002.

4.DISCUSSION

Since the second half of the 20th centuryin many places of Europe, changes of average phenological dates have been observed in both the world of wild-growing and crop plants (Ahas et al., 2002;Bonofiglio et al., 2009; Chmielewski et al., 2004; Menzel, 2000; Menzel & Estrella, 2001). The most frequently shown direction of change has been in the acceleration of phenological dates, especially in relation to spring and summer dates. The strongest acceleration of phenological dates has been observed since the end of the 1980s (Kalbarczyk, 2009a; Kalvāne et al., 2009; Schleip et al., 2009a). However, the size and direction of the changes of phenological dates is spatially diverse. In many research studies, acceleration of phenological dates has been confirmed mainly in the north-western part of the Europe, but also in Central Europe, in the Black Sea region, in the Baltic Sea region and around the Carpathians (Estrella et al., 2007; Kalvāne et al., 2009; Menzel & Estrella, 2001; Mozny et al., 2009). Delay of phenological dates according to Schleip et al. (2009a) has been observed in central Poland and in the Baltic Sea region. Delay of phenological dates, proved by research, mainly pertains to the autumn period, but in this case the direction of change is diverse, depending on the examined plant species (Gordo & Sanz, 2009).

TABLE 5. Statistical characteristics of the average air temperature in the growing season (May-October) of tomato, and in the periods significantly affecting agrotechnical dates and the dates of phenological phases in Poland, in the years 1965-2004.

Period	Temperature (°C)						Absolute temperature (°C)	
	mean	highest	lowest	range[1]	S	trend (°C/10a)	max	min
May	13.0	16.3	9.3	7.0	1.5	0.4***	18.7	7.5
10th May-10th June	14.3	18.0	11.1	6.9	1.8	n.s.	19.7	9.2
20th May-20th June	15.0	18.5	12.7	5.8	1.4	n.s.	21.3	10.6
10th August-10th September	16.1	19.3	13.5	5.8	1.1	0.3***	20.5	12.5
May-October	14.3	15.7	12.9	2.8	0.7	0.3**	18.0	11.6

max – absolute maximum temperature, min – absolute minimum temperature, [1]between the highest and the lowest temperature. The remaining explanations see Tables 2 and 3.

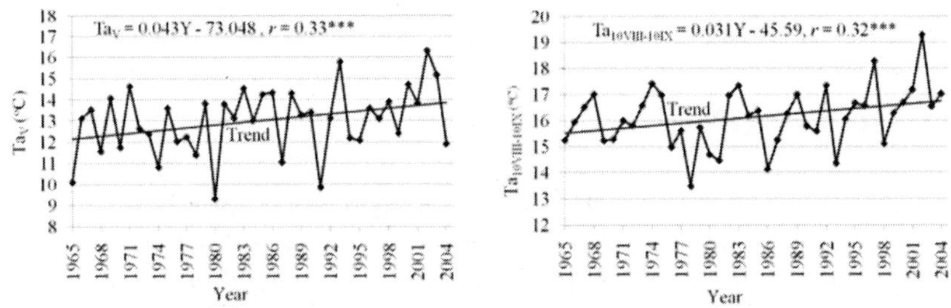

r - correlation coefficient, *** significant at $p \leq 0.01$.

FIGURE 11 Temporal distribution of the average air temperature in May (Tav) and in the period 10th August-10th September (Ta$_{10VIII-10IX}$) in Poland, in the subsequent years of the analysed multi-annual period 1965-2004.

The observed changes most often oscillate from 1 to several days per 10 years, and less frequently – about a dozen days per 10 years (Fujisawa & Kobayashi, 2010; Tao et al., 2006). Acceleration of hop flowering observed in the Czech Republic amounted to -1.6 days/10 years (Mozny et al., 2009). In Germany changes of certain phenological dates of winter rye and fruit trees amounted to -2.0 to -2.9 days/10 years (Chmielewski et al., 2004). According to Estrella et al. (2007) in the multi-annual period 1951-2004, average acceleration of phenological dates of field plants and vegetables in Germany amounted to -1.1 to -1.3 days /10 years. In the Mediterranean region, changes of leaf unfolding, flowering and fruiting which had been measured since the mid-70s amounted to -3.2 to -5.9 days /10 years(Gordo & Sanz, 2009).A much bigger acceleration occurred in the case of wild-growing trees; in Lithuania and Latvia, European hazel flowering was accelerated by -1 to -11 days /10 years and alder bloomed earlier from -1 to -15 days /10 years (Kalvāne et al., 2009).

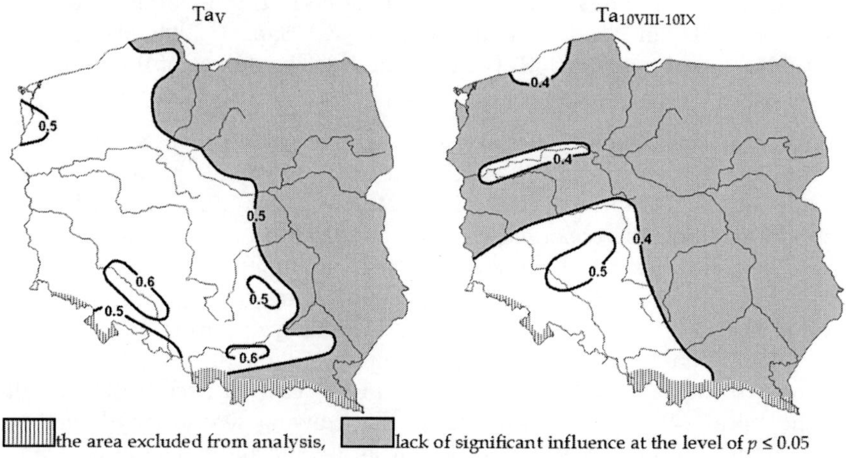

FIGURE 12. Statistically significant, at least at the level of $p \leq 0.05$, linear regression coefficients for the trend of the average air temperature in May (Tav) and in the period 10th August-10th September ($Ta_{10VIII-10IX}$) in Poland, calculated for 10 years, in the years 1965-2004.

During the 40-year period of the present research, there was an acceleration of almost all phenological dates of tomato. This acceleration amounted to -0.6 to -3.5 days /10 years. The biggest acceleration concerned the dates occurring at the end of June and the beginning of July. Acceleration of dates of similar sizewas observed in Poland in relation to other vegetal and crop plants (Kalbarczyk, 2009a,2009b). The results show that thermophilous vegetal plants cultivated in Poland are subjected to similar thermal conditions as plants cultivated in the western part of Europe. However, persistence of a negative trend for the tomato dates in June, over several years, can be evaluated as not very favourable. Cultivating tomatoes at too early a date is a risk because of the possibility of a late spring frost (Kalbarczyk, 2010b). In Poland the average date of occurrence of the last spring frost is marked by a negative trend (Koźmiński et al., 2010). However, such a trend was not confirmed for the absolute dates of frost occurrence. In Poland, the last spring frost may occur even at the end of June.

The above-mentioned possible delay of autumn phenological dates (Gordo & Sanz, 2009; Matsumoto, 2010), in the case of tomato cultivated in Poland, manifests itself by delay of the end of harvesting. In the examined multi-annual period, the end of harvesting was delayed by 2 days /10 years. The proved later course of some dates can be found in the research conducted in Spain, Japan and China (Gordo & Sanz, 2009; Matsumoto, 2010; Matsumoto et al., 2003; Tao et al., 2006). The degree of the observed delay was within a range of from 3.6 to as much as 21 days /10 years.

Changes of the phenological dates cause diverse duration of particular development stages of plants, and duration of the whole growing season (Liu et al., 2010; Moiseev et al., 2010; Mozny et al., 2009). According to research studies, the observed changes in duration of the growing period most frequently lead to lengthening, which can amount to about a dozen days (Kalvāne et al., 2009; Song et al., 2008). Lengthening of the growing period occurs in the case of an earlier beginning of spring and summer phenological dates and a later course of autumn phenophases. Lengthening of the growing season may even occur in the case of acceleration of all phenological dates.In the Baltic states, the growing season lengthened in spite of the recorded acceleration of both spring and autumn dates (Kalvāne et al., 2009). This lengthening was a consequence of a big dominance of spring acceleration over autumn acceleration. In the 31-year research period, the growing season of wild-growing trees in Lithuania and Latvia lengthened by 7 days (Kalvāne et al., 2009). Lengthening of the growing season by 8% (18 days) was also confirmed in the Mediterranean region (Gordo & Sanz, 2009). The cause of this lengthening was also mainly the acceleration of the spring phases.

The duration of the particular developmental stages of plants underwent more diverse changes. A year by year shortening in the duration of the spring and summer development stages is observed most frequently. However, extreme differences occur which depend on: development stage, kind of plant, and the area of its cultivation or occurrence. In Germany, significant shortening of the period from sowing to emergence of maize was found to be, on average 1.6 days/10 years (Chmielewski et al., 2004). However, duration of the remaining part of maize growth and development, i.e. from emergence to the beginning of harvesting, lengthened, on average, 2.1 days/10 years. Significant lengthening of development stages was also proved for the period from the beginning of stem elongation to the beginning of heading of winter rye, and the beginning of row closing to the beginning of harvest of sugar beet. The degree of lengthening amounted to 1.0 day/10 years for winter rye and 1.2 days/10 years for sugar beet (Chmielewski et al., 2004). In Poland, negative trends in the duration of some development stages were proved for, e.g., onion, medium-early potato and medium-late potato (Kalbarczyk, 2009b, Kalbarczyk E. & Kalbarczyk R., 2010). A significant shortening of duration for two onion periods was also proved. For the period from sowing to the end of emergence, shortening amounted to -1.7 day/10 years; for the period from the end of emergence to the beginning of leaf bending it amounted to -0.7 day/10 years. Similar results were obtained for potato. Medium-late potato cultivars were characterised by a shortening of the period from emergence to flowering by -1.7 days/10 years. For medium-early potato cultivars, shortening of the period from haulm drying to harvesting was -0.8 day/10 years. On the other hand, for both groups of potatoes, the period from flowering to haulm drying lengthened by about 2-3 days/10 years (Kalbarczyk E. & Kalbarczyk R., 2010). In the case of tomato, in the years 1965-2004, only the period Bfs-Bh became significantly shorter; on average by -1.4 days/10 years. However, an opposite tendency was seen in the case of the period from the beginning of harvesting to the end of harvesting of the plant. The period from the beginning of harvesting to the end of harvesting, on average, lengthened by as much as 5.6 days/10 years. Changes in the duration of particular tomato development stages caused the whole season, which lasts from planting up to the end of harvesting, to lengthen by 2.6 days/10 years. It seems that a longer period of tomato harvesting creates the possibility of achieving a bigger yield of the plant.

The conducted research confirmed that, in the case of tomato, like the majority of plants (Craufurd & Wheeler, 2009; Morin et al., 2010; Schliep et al., 2009b), changes in the course of tomato phenological dates and in duration of tomato development stages are significantly dependent on changes in air temperature. The relationship between tomato growth and development and temperature was confirmed in the research by Bojacá et al. (2009) and van der Ploeg & Heuvelink (2005). All phenological and agrotechnical dates of tomato considered in the present study, were significantly dependent on air temperature. The first three dates: Bf, Bfs, Bh, were negatively correlated with air temperature, and the last considered date, Eh, was positively correlated with air temperature. The degree of change in the tomato phenological date, caused by a 1°C temperature change, amounted to 1.8 to 3.1 days. Similar reactions were observed in other crop plants in Poland and Germany (Chmielewski et al., 2004; Kalbarczyk E. & Kalbarczyk R., 2010). A statistically confirmed air temperature increase in May and in the period from 10th August to 10th September amounted to, respectively, 0.4°C/10 years and 0.3°C/10 years. In southern Poland, the temperature trend in May amounted to as much as 0.6°C/10 years. The obtained results are similar to those described in different parts of Europe (Bauer et al., 2009; Bonofiglio et al., 2009; Chmielewski et al., 2004; Kapur et al., 2007; Saue & Kadaja, 2010) and the world (Lobell et al., 2007; Song et al., 2008; Wang et al., 2008).

Air temperature may affect the quantity and quality of the yield directly and indirectly (Tshiala&Olwoch, 2010). Indirect influence of air temperature is seen through the proven effect on the rate of growth and development of plants. The present work confirms changes occurring in the course of tomato development in Poland. However, there are no clear observations corroborating an increase or decrease in tomato yield occurring from this cause.

5.CONCLUSIONS

An increase, by 0.3°C /10 years, in average air temperature in the growing season of tomato (May-October), contributed to the changes of not only the dates of phenological phases and development stages but also to the agrotechnical dates of the plant. In the years 1965-2004, important changes in temporal distribution of all the considered dates were noted, and of the period duration from planting up to the end of harvesting. In Poland, the growing season of field-cultivated tomato lengthened by 2.6days /10 years, which resulted both from acceleration of the dates of: planting up (-0.6 day /10 years), flowering (-0.7 days /10 years), fruit-setting (-1.1 days /10 years), the beginning of harvesting (-3.5 days /10 years), and delay of the end of harvesting date (2.1 days /10 years). The biggest changes in the development of the tomato were recorded in the fruiting period - from the beginning to the end of harvesting, which for the whole country, on average, lengthened by 5.6 days / 10 years. However, in Poland this lengthening was not even, as it fluctuated from less than 1.5 days /10 years in the northern, north-eastern and central-eastern parts of the country, to above 2.5 days /10 years in the southern part.

An increase, by 1°C, in the average air temperature during the periods which significantly affect the dates of tomato caused acceleration of: the beginning of flowering by 1.8 days, the beginning of fruit-setting by 2.2 days, the beginning of harvesting by 3.1 days and delay of the end of harvesting by 2.1 days. During the 1965-2004, 40-year research period, deviations from the multi-annual average of the agrotechnical dates and duration of tomato development stages were discovered. These deviations were a result of the change in thermal air conditions. The date of the beginning of harvesting differed most from the average date in 1965, and the date of the end harvesting differed most from the average date, in the years: 1978, 1997 and 2002. Positive deviations from the norm of agrotechnical dates in Poland, oscillated from less than 4 to more than 8 days, in the years 1965 and 2002, and from less than 2 to more than 6 days in 1997. Negative deviations from the norm of agrotechnical dates in Poland, oscillated from more than 6 to less than 8 days, in 1978. The length of the period from the beginning of harvesting to the end of harvesting differed most from the multi-annual average in 2002, and the length of the period from planting up to the end of harvesting in 1978 and 2002. The highest deviation from the norm, for the duration of the period from the beginning to the end of harvesting, amounted to more than 14 days, and occurred in the central-west and in the Warsaw region.

Changes in the course of tomato development found in the present work do not lead to evident changes in cropping of this plant. However many things indicate, that the occurring changes may improve the conditions of tomato field-cultivation in Poland, in the future.According to the IPCC report (2007), it is predicted that as a result of climate change the yield of crop plants in central Poland will decrease on average by 5%. On the other hand, there are forecasts of an increase in the yield in the north by 5% and by 30% in the submontaineous regions. The changes in the yield quantity will vary depending on a type of plants. The highest yield increase, by several dozen per cent, is predicted for thermophilous plants, which will result from increased thermal resources.

REFERENCES

1. R. Ahas, A. Aasa, A. Menzel, V. G. Fedotova, H. Scheifinger, 2002 Changes in European spring phenology International Journal of Climatology 1727 1738
2. M. Ahmed, A. Hamid, Z. Akbar, 2004Growth and yield performance of six cucumber (Cucumis sativus L.) cultivars under agro-climatic conditions of Rawalakot, Azad Jammu and Kashmir. International Journal of Agriculture and Biology, 6, 396 399
3. J. . Babik, Ed., 2004 Ecological methods of tomato cultivationin the ground and under cover. Krajowe Centrum Rolnictwa Ekologicznego, 8-38906-036-1Poland [In Polish]
4. Z. Bauer, M. Trnka, J. Bauerová, M. Možný, P. Štěpánek, L. Bartošová, Z. Žalud, 2009Changing climate and the phenological response of great tit and collared flycatcher populations in floodplain forest ecosystems in Central Europe. International Journal of Biometeorology, 54, 99 111
5. C. R. Bojacá, R. Gil, A. Cooman, 2009 Use of geostatistical and crop growth modelling to assess the variability of greenhouse tomato

yield caused by spatial temperature variations Computers and electronics in agriculture 219 227

6. T. Bonofiglio, F. Orlandi, C. Sgromo, B. Romano, M. Fornaciari, 2009 Evidences of olive pollination date variations in relation to spring temperature trends Aerobiologia 227 237
7. F. Chmielewski, M. , A. Müller, E. Bruns, 2004 Climate changes and trends in phenology of fruit trees and fields crops in Germany, 1961-200 Agricultural and Forest Meteorology 69 78
8. F. M. Chmielewski, A. Müller, W. Küchler, 2005 Possible impacts of climate change on natural vegetation in Saxony (Germany) International Journal of Biometeorology 96 104
9. P. Q. Craufurd, T. R. Wheeler, 2009 Climate change and the flowering time of annual crops Journal of Experimental Botany 2529 2539
10. N. R. Dalezios, A. Loukas, D. Bampzelis, 2002 The role of agrometeorological and agrohydrological indices in the phenology of wheat in central Greece Physics and Chemistry of the Earth, 27, 1019 1023
11. M. Dobosz, 2001 Computerized statistical analysis of results (edition 1), EXIT Publisher, 838767429Warszawa, Poland [in Polish]
12. P. . Domański, Ed., 1998 Methodology of research on economic value of field-grown plants varieties. Solanaceous vegetables. COBORU, Słupia Wielka, Poland [in Polish]
13. N. Estrella, T. Sparks, A. Menzel, 2007 Trends and temperature response in the phenology of crops in Germany Global Change Biology, 13, 1737 1747
14. FAO, 2008Available from: http\\www. faostat.fao.org
15. M. Fujisawa, K. Kobayashi, 2010 Apple (Malus pumila var. domestica) phenology is advancing due to rising air temperature in northern Japan. Global Change Biology, 16, 2651 2660
16. O. Gordo, J. J. Sanz, 2009 Long-term temporal changes of plant phenology in the Western Mediterranean Global Change Biology 1930 1948
17. GUS, 2010Statistical Yearbook of Agriculture 2010. Available from: http\\www.stat.gov.pl
18. IPCC, 2007Fourth Assessment Report- Climate Change 2007: Synthesis Report 2007

19. E. Jorquera-Fontena, R. Orrego-Verdugo, 2010Impact of global warming on the phenology of a variety of grapevine grown in southern Chile. Agrociencia 427 435
20. R. Kalbarczyk, 2009a Air temperature changes and phenological phases of field cucumber (Cucumis sativus L.) in Poland, 1966-2005 Horticultural Science 75 83
21. R. Kalbarczyk, 2009bThe effect of climate change in Poland on the phenological phases of onion (Allium cepa L.) between 1966 and 2005.Agriculturae Conspectus Scientificus, 74(4), 297 304
22. R. Kalbarczyk, 2010a Unfavourable thermal conditions of air at the turn of the 20th and 21st centuries reducing crop productivity of pickling cucumber (Cucumis sativus L.) in Poland Spanish Journal of Agricultural Research, 8(4), 1163 1173
23. R. Kalbarczyk, 2010bSpatial and temporal variability of the occurrence of ground frost in Poland and its effect on growth, development and yield of pickling cucumber (Cucumis sativus L.), 1966-2005. Acta Scientiarum PolonorumHortorum Cultus, 9(3), 3 26
24. E. Kalbarczyk, R. Kalbarczyk, 2010The course phenological phases of potato and its determination by multi-annual variability of air temperature in Poland. Annales UMCS Sectio E, 4, 1 11in Polish, with English abstract]
25. G. Kalvāne, D. Romanovskaja, A. Briede, E. Bakšiene, 2009Influence of climate change on phonological phases in Latvia and Lithuania. Climate Research, 39, 209 219
26. B. Kapur, P. Steduto, M. Todorovic, 2007 Prediction of climatic change for the next 100 years in the Apulia Region, Southern Italy The Italian Journal of Agronomy / Rivista di Agronomia 4, 365 371
27. C. Koźmiński, B. Michalska, J. Leśny, 2010 Climate risk in agriculture in Poland, (edition 1), Uniwersytet Szczeciński Publisher, 978-8-37241-743-5Szczecin, Poland [in Polish, with English abstract]
28. Z. W. Kundzewicz, P. Kowalczak, 2008Climate changes and its effects(edition 1), Kurpisz S.A. Publisher, 978-8-37524-969-9Poznań, Poland [in Polish]
29. B. Liu, M. Henderson, Y. Zhang, M. Xu, 2010 Spatiotemporal change in China's climatic growing season: 1955-2000 Climatic Change 93 118

30. D. B. Lobell, K. N. Cahill, C. B. Field, 2007 Historical effect of temperature and precipitation on California crop yields. Climatic Change 187 203
31. K. Matsumoto, 2010 Causal factors for spatial variation in long-term phonological trends in Ginkgo biloba L. in Japan International Journal of Climatology 1280 1288
32. K. Matsumoto, T. Ohta, M. Irasawa, T. Nakamura, 2003Climate change and extension of the Ginko biloba L. growing season in Japan. Global Change Biology, 9, 1634 1642
33. W. Mazurczyk, B. Lutomirska, A. Wierzbicka, 2003 Relation between air temperature and length of vegetation period of potato crops Agricultural and Forest Meteorology 169 172
34. U. . Meier, Ed., 2001BBCH [Monograph]. Growth Stages of Mono- and Dicotyledonous Plants. Federal Biological Research Centre for Agriculture and Forestry, Berlin, 130 133
35. A. Menzel, 2000 Trends in phenological phases in Europe between 1951 and 1996. International Journal of Biometeorology 76 81
36. A. Menzel, N. Estrella, 2001 Plant phenological changes. In: Fingerprints of climate change: adapted behaviour and shifting species ranges. G.R., Walther; C.A., Burga & P.J. Edwards (Eds.), 123 137Kluwer/Plenum, New York, U.S.A.
37. P. A. Moiseev, A. A. Bartysh, Z. Y. Nagimov, 2010 Climate changes and tree stand dynamics at the upper limit of their growth in the north Ural Mountains Russian Journal of Ecology 486 497
38. X. Morin, J. Roy, L. Sonie, I. Chuine, 2010 Changes in leaf phenology of three European oak species in response to experimental climate change New Phytologist 900 910
39. M. Mozny, R. Tolasz, J. Nekova,r, T. Sparks, M. Trnka, Z. Zalud, 2009 The impact of climate change on the yield and quality of Saaz hops in the Czech Republic Agricultural and Forest Meteorology, 149, 913 919
40. S. Parey, 2008 Extremely high temperatures in France at the end of the century ClimateDynamics, 30, 99 112
41. S. Peng, J. Huang, J. E. Sheehy, R. C. Laza, R. M. Visperas, X. Zhong, G. S. Centeno, G. S. Khush, K. G. Cassman, 2004Rice yields ecline wih higher night temperature from global warming. Proceedings of the National Academy of ScienceUSA, 101(27), 9971 9975

42. D. R. Peiris, J. W. Crawford, C. Grashoff, R. A. Jefferies, J. R. . Porter, B. Marshall, 1996A simulation study of crop growth anddevelopment under climate change. Agricultural and Forest Meteorology,79, 271 287
43. H. Pokladníková, J. Rožnovský, T. Středa, 2008Evaluation of soil temperatures at agroclimatological station Pohořelice. Soil & Water Research, 3(4), 223 230
44. E. G. Popov, A. V. Talanov, V. K. Kurets, S. N. Drozdov, 2003 Effect of temperature on diurnal changes in CO2 exchange in intact cucumber plants Russian Journal of Plant Physiology 178 182
45. Saue, T. & Kadaja, J. (2010). Simulated potato crop yield as an indicator of climate variability and changes in Estonia. In: Climate Change and Variability, S. Simard (Ed.), InTech, pp. 366-388, ISBN 978-953-307-144-2, Available from http://www.intechweb.org/search?q=Climate+Change+and+Variability
46. C. Schleip, T. C. Sparks, N. Estrella, A. Menzel, 2009a Spatial variation in onset dates and trends in phenology across Europe Climate Research, 39, 249 260
47. C. Schleip, A. Rais, A. Menzel, 2009b Bayesian analysis of temperature sensitivity of plant phenology in Germany Agricultural and Forest Meteorology 1699 1708
48. H. Skąpski, A. Borowy, 2000Solanaceous vegetables. Tomato, In: Field cultivation of vegetables, M. Orłowski, (Ed.), 170 190Brasika Publisher, 8-39028-215-1Poland [in Polish]
49. M. Sobczyk, 1998 Statistics.Theoretical grounds, examples-problems (edition 1),University of Maria Curie-Skłodowska Publisher, 8-32271-153-0Poland [in Polish]
50. Y. Song, H. W. Linderholm, D. Chen, A. Walther, 2008 Trends of the thermal growing season in China, 1951-2007 International Journal of Climatology 33 43
51. M. I. Sysoeva, E. F. Markovskaya, T. G. Kharkina, 1997 Optimal temperature drop for the growth and development of young cucumber plants Plant Growth Regulation 135 139
52. F. Tao, M. Yokozawa, Y. Xu, Y. Hayashi, Z. Zhang, 2006 Climate changes and trends in phenology and yields of field crops in China, 1981-2000 Agricultural Forest Meteorology, 138, 82 92

53. M. F. Tshiala, J. M. Olwoch, 2010Impact of climate variability on tomato production inLimpopo Province, South Africa.African Journal of Agricultural Research, 5(21), 2945 2951
54. H. L. Wang, Y. T. Gan, R. Y. Wang, J. Y. Niu, H. Zhao, Q. G. Yang, G. C. Li, 2008Phenological trends in winter wheat and spring cotton in response to climate changes in northwestern China. Agricultural Forest Meteorology, 148, 1242 1251
55. Y. P. Wang, D. J. Connor, 1996Simulation of optimal development for spring wheat at two locations in southern Australia under present and changed climate conditions. Agricultural Forest Meteorology, 79, 9 28
56. A. Węgrzyn, 2007Classification of vegetation periods according to thermal criteria in the Lublin region in the years 1951-1990. Acta Agrophysica, 9(2), 505 516In Polish, with English abstract]
57. J. Wolf, M. van Oijen, 2002Modelling the dependence of European potato yields on changes in climate and CO2. Agricultural and Forest Meteorology, 112, 217 231
58. A. Van der Ploeg, E. Heuvelink, 2005Influence of sub-optimal temperature on tomato growth and yield: a review. Journal of Horticultural Science & Biotechnology, 80(6), 652 659
59. G. Xiao, Q. Zhang, Y. Yao, H. Zhao, R. Wang, H. Bai, F. Zhang, 2008Impact of recent climatic change on the yield of winter wheat at low and high altitudes in semi-arid northwestern China. Agriculture, Ecosystems & Environment, 127, 37 42
60. F. Yoshie, 2010Vegetative phenology of alpine plants at Tateyama Murodo-daira in central Japan. Journal of Plant Research, 123, 675 688
61. E. Żmudzka, 2004The climatic background of agricultural production in Poland (1951-2000). Miscellanea Geographica, 11, 127 137

8

CHAPTER

REMOTE SENSING OF IRRIGATED AGRICULTURE: OPPORTUNITIES AND CHALLENGES

Mutlu Ozdogan *, Yang Yang, George Allez and Chelsea Cervantes

Center for Sustainability and the Global Environment (SAGE), University of Wisconsin-Madison, 1710 University Avenue, Madison, WI 53726, USA

ABSTRACT

Over the last several decades, remote sensing has emerged as an effective tool to monitor irrigated lands over a variety of climatic conditions and locations. The objective of this review, which summarizes the methods and the results of existing remote sensing studies, is to synthesize principle findings and assess the state of the art. We take a taxonomic approach to group studies based on location, scale, inputs, and methods, in an effort to categorize different approaches within a logical framework. We seek to evaluate the ability of remote sensing to provide synoptic and timely coverage of irrigated lands in several spectral regions. We also investigate the value of archived data that enable comparison of images through time. This overview of the studies to date indicates that remote sensing-based monitoring of irrigation is at an intermediate stage of development at local scales. For instance, there is overwhelming consensus on the efficacy of vegetation indices in identifying irrigated fields. Also, single date imagery, acquired at peak growing season, may suffice to identify irrigated lands, although to multi-date image data are necessary for improved classification and to distinguish different crop types. At local scales, the mapping of irrigated lands with remote sensing is also strongly affected by the timing of image acquisition and the number of images used. At the regional and global scales, on the other hand, remote sensing has not been fully operational, as methods that work in one place and time are not necessarily transferable to other locations and periods. Thus, at larger scales, more work is required to indentify the best spectral indices, best time periods, and best classification methods under different climatological and cultural environments. Existing studies at regional scales also establish the fact that both remote sensing and national statistical approaches require further refinement with a substantial investment of time and resources for ground-truthing. An additional challenge in mapping irrigation across large areas occurs in fragmented landscapes with small irrigated and cultivated fields, where the spatial scale of observations is pitted against the need for high frequency temporal acquisitions. Finally, this review identifies passive and active microwave observations, advanced image classification methods, and data fusion including optical and radar sensors or with information from sources with multiple spatial and temporal characteristics as key areas where additional research is needed.

Keywords: irrigation; agriculture; remote sensing; image classification; resolution

1. INTRODUCTION

The intensification of agricultural practices—under the auspices of the "Green Revolution" that includes better seeds, extensive fertilizer use, and irrigation—has dramatically altered the relationship between humans and environmental systems across the world. Today many agricultural lands are being used much more intensively as opportunities for expansion are being exhausted elsewhere. In the last 40 years, global agricultural production has more than doubled—although cropland has increased by only 12%—in part through increased reliance on irrigation [1,2]. Currently, irrigated agriculture is the principal consumer of fresh water resources; it accounts for more than 70 percent of water withdrawn from lakes, rivers, and groundwater aquifers [3]. As the earth's population continues to increase and the demand for food, fuel, and fiber rises, continued agricultural intensification will require at least a 50 percent increase in water resources, especially in arid and semi-arid regions [4].

While these modern agricultural practices have successfully increased food production, they have also caused significant environmental change in many regions. Accurate information on the extent of irrigation is thus fundamental to many aspects of Earth System Science, and global change research in general. These aspects include modeling of water exchange between the land surface and atmosphere [5,6,7,8], analysis of the impact of climate change and variability on irrigation water requirements and supply [9,10,11,12,13], management of water resources that affect global food security [14], and climatic feedbacks, including the effect that results from evaporative cooling in intensely irrigated arid areas [15,16].

Despite their significance for food security and the water and energy cycles, the extent and distribution of irrigated areas worldwide still remain uncertain [17]. Existing maps, especially those covering large areas, have been derived primarily from country-level statistics. The politically charged nature of irrigation often sets the stage for under-reporting of water use; this is especially true in countries that share resources across borders with their neighbors [18]. Country-level estimates also mask the considerable spatial variability in irrigation practices, and simply cannot reflect the location or extent of irrigation across large areas [19]. Even in countries such as the U.S., where the extent of irrigated areas is known, irrigation-related information exists in disparate sources and cannot be easily synthesized into a single continental scale database [20]. Also, information on irrigated areas in many countries is reported only from officially recognized management units (or command areas) serviced by large scale irrigation projects. As a result, subsistence-scale irrigation is not reported, although these areas could collectively account for a substantial land area and significant amount of water use.

Satellite remote sensing offers tremendous potential for routine monitoring of irrigation due to the synoptic nature of the data and readily available archives of imagery. Yet studies that have used remote sensing to map irrigated lands remain relatively rare. This is a direct result of the complexity associated with trying to map *land use* as opposed to *land cover*. While it may be straightforward to detect the high near-infrared signal of mature crops given appropriate spatial, spectral and temporal resolution data (*i.e.*, land *cover*), detecting irrigation requires knowledge of land management, or some understanding of where and when humans have provided water or supplemented rain-fed crops (*i.e.*, land *use*). Because of the difficulty in isolating these practices with satellite observations, a literature search reveals only 65 peer-reviewed papers that use remote sensing to map irrigation, compared to thousands that report agricultural or land-cover mapping activities. Thus, from the remote sensing perspective, studies that attempt to map irrigated areas have been rare and scientific consensus on mapping methodologies is fragmented and evolving.

This review seeks to synthesize current studies on identification and mapping of irrigated areas by remote sensing. Our goal is two-fold. First, we will provide a reference guide to the spatial, spectral, and temporal information requirements for monitoring irrigated areas, derived from case studies that have successfully mapped irrigated lands. Second, and more important, we will establish the state-of-the-art in this field by providing a comprehensive assessment and a taxonomic synthesis of studies to date. This information can provide a foundation for future studies to expand on these methods and fill data gaps. The approaches that have been adopted to tackle irrigation are diverse; they vary in scale, extent, data inputs and processing requirements. It is also clear that a consensus within the scientific community as to the 'best practices' for mapping irrigation are still evolving, although certain methods appear to be common among different studies. Moreover, reviews such as this portray information needs for timely and accurate monitoring of irrigation. This is necessary in order to form the basis for development of sustainable water management practices within the context of what is perhaps the greatest human intervention in the hydrological cycle.

Advantages and disadvantages of remote sensing

We will begin with a brief discussion of the benefits and drawbacks of remote sensing for mapping crop location, productivity, and change in irrigated settings. Remote sensing has been an effective tool to monitor irrigated lands in many locations around the world under a variety of environmental conditions [19,20,21,28,33,35,50,60,65]. It provides synoptic coverage of irrigated fields in several spectral regions and with temporal frequencies sufficient to assess vegetation growth, maturity, and harvest. Archived data that span many years allow comparison of images, thus revealing change. The digital nature of satellite data also makes it relatively easy to integrate into a Geographic Information System (GIS) for synthesis or comparison with other data sources. Remotely sensed data are also less costly and time-consuming than traditional statistical surveys that may require aerial photography over large areas. This makes remote sensing particularly valuable for inventories of irrigated land and for monitoring in developing countries, where funds are limited and little objective information is available. Moreover, remote sensing delivers useful spatial information on the exact locations of irrigated lands rather than mere totals within arbitrary political units. This is important for prioritizing water delivery, assessing irrigation performance, providing irrigation intensities (e.g., single crop *vs.* double crop), quantifying environmental impact, objectively assessing irrigation water use and understanding changes where irrigation occurs. Finally, remote sensing can provide information on timing, both in the number of irrigation-related vegetation peaks and in the length of time irrigation is utilized over the course of a year.

However, satellite imagery also has limitations. Because of the spatial resolution of most operational imagery (15–60 m), it is difficult to identify small irrigated areas which, taken together, may cover significant parts of the earth. It is also difficult to separate irrigated fields from non-irrigated plots in humid areas because of substantial overlap in their spectral signatures. For example, the signatures of flooded irrigated fields at certain growth stages may overlap with those of natural wetlands, thus limiting accuracy in mapping. Researchers have overcome these limitations by using temporal information on crop planting, maturity, and harvest in conjunction with spectral information [21]. Unfortunately, the collection of remotely-sensed data is fixed by a given satellite's orbit and return interval, and thus observations are not always captured at ideal times (e.g., green-up or harvest). Optical data availability is also problematic in areas with frequent cloud cover, such as humid tropical and sub-tropical environments.

Having stated this, it is important to point out the technological advances made in remote sensing. For example, satellite constellations such as Rapideye with 5 meter spatial resolution and providing data in five spectral bands has already covered nearly 95% of USA geographic area in less than one year after launch. Further, multi-sensor data fusion (e.g., IRS, Rapideye, Landsat) are becoming increasingly important and feasible. Finally, looking at the limitations of conventional datasets such as subjectivity in data collection and varying statistical design in different studies, limitations from remote sensing by itself are less certain.

The final limitation considered here comes from the fact that identification of agricultural fields using remote sensing is difficult because irrigated landscapes are a subclass of croplands that themselves have traditionally been difficult to map [22,23,24]. Agricultural fields (and especially irrigated fields) are highly dynamic because each field may be at a different stage of development, and thus subject to being confused with natural land cover classes. Accuracy of land-cover maps is often inversely related to their categorical detail. Since agriculture is already inherently difficult to identify and map, the task of identifying irrigated areas as a subclass of cultivation becomes even more difficult. Perhaps this is where temporal data profiles will be invaluable to separate irrigation from rainfed agriculture as successfully demonstrated by [21,61]. Moreover, ancillary datasets on precipitation and evapotranspiration will come in handy when interpreting these temporal profiles [20].

Definition of irrigation

If we are to identify and map irrigation with remote sensing, a precise definition of what is considered to be irrigated is needed. In this review, we define irrigated lands as areas that receive full or partial application of water by artificial means to offset periods of precipitation shortfalls during the growing period. Fully irrigated areas are those where more than 60 percent of crop water requirements are met artificially; partially irrigated lands (or supplementally irrigated areas) receive between 30 and 60 percent artificially. Note that both surface and groundwater deliveries are included in this definition provided human intervention to move water from one location to another is involved.

Irrigation is practiced in virtually every country, at scales ranging from subsistence farming to national enterprise. The precise location is determined by a combination of factors that include climate, resource availability, crop patterns, and technical expertise. Climate plays an important role in the distribution of irrigation as it determines natural moisture availability (precipitation), crop demand (evaporation), and crop schedules. In humid climates, irrigation often takes the form of a supplemental water supply to meet the excess demand of crops whose growth cycle may be out of sync with natural precipitation. In arid and semi-arid climates, continual irrigation is often necessary to assure agricultural production.

While climate is an important driver of the need for irrigation, it is water availability that primarily determines its existence and sustainability. Currently, groundwater is by far the greater source for irrigation [25]. In regions where withdrawals for agriculture exceed recharge rates, the quantity and quality of groundwater quickly deteriorates, jeopardizing its sustainability. Thus nations may be forced to decide between agricultural and domestic use in their allocation of groundwater. Surface water for irrigation purposes appears more sustainable, but this is deceptive. It requires large structures involving complex engineering such as dams, conveyers, and canals to redistribute the resources. Furthermore, most river basins span international boundaries. Diversions of water for irrigation upstream often reduce its availability downstream, leading to international conflicts as in the case in the Middle East and Central Asia.

2. REVIEW OF EXISTING STUDIES

In this section, we review existing studies on irrigation mapping with remote sensing and assess data and methodological features that are common and practical. We have chosen spatial scale to categorize these studies. Here *spatial scale* identifies the scope of the study area and is defined as *local, regional,* or *global*. Local studies refer to one or more irrigation basins or command areas. Regional studies include large river basins and continental areas, while global studies present attempts to map irrigation worldwide. This conceptual framework is helpful both to understand the processes involved in each category and to classify mapping approaches. Within each category, the discussion is further organized around the nature of remote sensing imagery and methods of processing these images. In tabular format we describe the advantages and disadvantages of satellite sensors that have been used to identify and map irrigated lands for the following imaging systems: Landsat, Satellite Pour l'Observation de la Terre (SPOT), China-Brazil Earth Resources Satellite (CBERS), Advanced Very High Resolution Radiometer (AVHRR), Moderate Resolution Imaging Spectroradiometer (MODIS), Medium Resolution Imaging Spectrometer (MERIS), and Indian Remote Sensing Satellite (IRS) (Table 1). Finally, we provide selected examples from the literature for image classification techniques to determine the most successful options in identifying irrigated areas and separating them from other land cover types (Table 2). The accuracy of these classifications is included to give an idea of each method's success.

Table 1. Current operational optical sensors and their technical/logistical capacities for irrigation mapping. Some of the table headings are adapted from the United Nations Collaborative Programme on Reducing Emissions from Deforestation (UN-REDD) sourcebook and Forest Degradation in Developing Countries [97].

Satellite observation system/program	Technical observation challenges solved	Access to information/ data worldwide	Continuous observation program with global coverage	Pre-processed datasets accessible	Image data cost	Technical difficulty required to produce maps	Frequency of use in irrigation studies
RapidEYE	yes	no	yes	yes	high	medium	low
Landsat	yes	yes	yes	yes	low	medium	high
SPOT	yes	yes	yes	yes	high	medium	medium
AWiFS	yes	no	no	yes	high	medium	low
LISS	yes	no	no	yes	high	medium	low
ASTER	yes	yes	no	yes	low	medium	low
CBERS	no	no	no	no	low	medium	low
THEOS	yes	no	no	yes	medium	medium	low
MODIS	yes	yes	yes	yes	low	high	medium
MERIS	yes	yes	yes	yes	low	high	low
AVHRR	yes	yes	yes	yes	low	high	medium
SPOT VEG.	yes	no/yes	yes	maybe	low	medium	low

Table 2. Summary of spatial scales, sensors, methods, and example applications in the literature.

Spatial scale	Sensors used	Method of mapping	References
local	Landsat TM/ETM+, SPOT, LISS, ASTER, AWiFS, CBERS, THEOS	Photo interpretation, Image arithmetic, Image classification, segmentation, image fusion	[26-29,31-33, 42-44,98]
regional	Landsat TM/ETM+,MODIS, MERIS, AVHRR, SPOT VGT	Times-series analysis, Supervised/unsupervised classification, masking	[21,47-50,53]
continental	Landsat TM/ETM+, MODIS, MERIS, AVHRR, SPOT VGT	Times-series analysis with other ancillary data, data fusion	[7,19-21,49, 52,98]
global	MODIS, MERIS, AVHRR, SPOT VGT	Unsupervised clustering, machine learning algorithms applied to time-series data, also employ other ancillary data(statistic, ground truth data…)	[22,54-57]

2.1. Local Scale Studies

To date, most research on remote sensing-based irrigation mapping occurs at local scales for several reasons. First, local areas provide familiar environments where investigators have extensive knowledge of their study sites. Second, the purpose of a remote sensing study may be to demonstrate the utility of a method rather than an operational assessment encompassing all irrigation. In these cases a small area suffices. Third, irrigation practices take many forms and this variation increases as the study area becomes larger. Therefore methods developed in one place and time may not be appropriate for other locations and durations.

Two forms of mapping methods emerge as most common in local area studies. These methods are visual interpretation and digital image classification. Early work concentrated on determining the utility of remotely sensed imagery using various visual interpretation techniques adapted from the airphoto sector to map and update irrigated acreage, mainly in the U.S. and India [26,27,28,29,30,31,32]. More recent studies have developed digital image classification methods specifically adapted to mapping irrigated lands automatically [33,34,35,36,37]. Within each category, several methods have been tested.

Visual interpretation of satellite imagery
Early work with satellite imagery at the local scale relied on visual interpretation of large hard-copy satellite image prints to identify and map irrigated lands [26,27,28]. These investigations recognized that satellite data provided a cost-effective method of obtaining multiple images during the growing season, an important consideration for identifying irrigation. Initially, traditional photo interpretation methods were used on Landsat

color composite images from multiple dates to draw boundaries of irrigated fields by hand. For visual interpretation, particular attributes of satellite data that proved useful were spectral availability in several channels including near-infrared, and temporal availability of imagery at low cost. Archival image data from multiple years also proved useful because newly irrigated croplands showed a color signature distinct from those previously irrigated. Since manual interpretation of satellite images is performed by an analyst, it is possible to use shapes of irrigated fields as an additional feature for identification [32]. Irrigated acreage inventories based on identifying and mapping the characteristic circular pivot system as seen in photographic enlargements of Landsat imagery are prime examples of these shape-based interpretation methods in the U.S. While this form of detection is simplistic, it provides an excellent example of the practical utility of remote sensing.

Visual interpretation of satellite data also includes automated approaches [29,30,38,39]. These studies benefit from the strong spectral separation of irrigated fields from harvested and fallow fields in the visible and near-infrared portions of the electromagnetic spectrum. One study [29] even demonstrated that in semiarid areas, croplands irrigated by surface water could be distinguished from those irrigated with ground water by visual interpretation of single-date Landsat imagery, although it was harder to identify different crop types.

Studies involving visual interpretation of satellite data also benefit from radiometric enhancements and manipulations in the form of spectral indices. Studies that compare these data to automated techniques of image classification report superior performance of manual mapping of area estimates when compared to reported acreage [30]. For example, by visually interpreting satellite imagery, Rao and Mohankumar [40] conducted an inventory of the cropland in a command area while evaluating various vegetation indices derived from Landsat-TM in India. The accuracy of the estimated irrigated crop area was within five percent of the estimate for the entire command area made by the Directorate of Economics and Statistics (DES) of the Indian government. Studies like this demonstrate that remote sensing is an effective time-saving technique to provide seasonal assessment of irrigated croplands and that the per hectare cost using remote sensing is often between one-half and one-fifth the cost of a conventional system based on air-photo interpretation supplemented by ground surveys.

While considered simplistic by today's image processing standards, the early work with visual interpretation of satellite data established several facts that hold true today. First, visual interpretation by analysts is by far the most accurate method of mapping land cover, irrigation included. This is because the human brain and expertise are the best image processors. However, visual interpretation can be expensive, although it is less costly than traditional air photo interpretations of large areas. Second, the red and the NIR portions of the electromagnetic spectrum provide tremendous information for recognizing irrigation. Third, multiple images during a single growing season increase accuracy because of differential growth stage times. Finally, while irrigation identification is possible with visual interpretation, identification of crop types is less reliable unless multiple images are available from the same growing season.

Digital image classification

Although visual interpretation of satellite images is useful, more recent work has emphasized digital image classification, partly because the analysis times are shorter and costs associated with mapping are lower. Common methods include multi-stage classification [31,41], unsupervised clustering [31,33,42], density slicing with thresholds [7,43,44], and decision tree classifications [45].

The multi-stage approach involves identification of land cover at increasingly refined categorical levels following the concept that irrigated areas are a subclass of cultivated lands, which themselves belong to vegetated landscapes. As noted, the accuracy of land-cover maps is often inversely related to their categorical detail. Thus, categorizing the landscape at each stage into more refined classes enables better identification of irrigated areas as a subclass.

As in visual interpretation, digital image classification benefits from spectral transformations [7,33,35,42,44]. In particular, the Normalized Difference Vegetation Index (NDVI) proves to be indispensible for identifying irrigated areas in local scale studies, in part because of differential spectral response between irrigated and non-irrigated fields. Common use of NDVI would include direct insertion into a classification algorithm as an input feature. Thresholds are set to distinguish productive fields from non-productive and non-irrigated areas, and also multi-temporal NDVI features such as the range and maximum NDVI. When multiple NDVI observations are available from the same growing season for several years, it is also possible to identify crop types as well as the changes in irrigated areas.

Multi-temporal data availability has also proved to be very useful in identifying irrigated areas. When using data from multiple time periods, the discrimination procedure is based on the different spectral responses of crops according to their phenological evolution. A number of studies have demonstrated that using spectral information from two consecutive seasons in a crop-year is sufficient to identify the irrigated crops, especially when only a few are dominant. However, for each season, the estimates require multiple images [36]. This is because single-date analysis in visible cropping intensity often does not take into account planting dates that vary from year to year. Therefore, multi-temporal analysis has greater potential to define irrigated areas [46]. Ultimately, classification results are conditional upon the temporal and spatial variability of the spectral signature of the land cover type in question, so suitable images must be available for the temporal approach to provide a complete inventory of all irrigated fields in a study area.

It is also possible to map irrigation with a single image during the peak irrigation season as long as its peak can be identified using a time-series of observations. One technique is to use coarse spatial resolution time series data to select the ideal time period where irrigation information is maximized and then acquire high resolution data to do the actual mapping at local and regional scales [7].

The review of local scale studies suggests that both methods baed on both visual intepretation and statistical and non-parametric classifications of image data work equally well in identifying irrigated areas. In general accuracies better tht 85 percent hav been achived with both approaches. In terms of spectral bands of importance, near-infrared combined with at least one visible light containing band have consistently produced higher classification accuracies than mapping applications that do not include these bands. It is also possible to map irrigation using single image NDVI data as long as the study area contains only a few classes to be distinguished. For example, these forms of classifications were more successful in mono cropped areas like the US mid-west when compared with fragmented irrigated landscapes of Asia and were quite futile in highly mixed cropping in Africa.

These studies also indicate that multi-temporal satellite remote sensing is a viable approach for identifying and monitoring irrigation. Methods based on multi-temporal NDVI features derived from Landsat data were particularly successful in distinguishing healthy irrigated lands from uncultivated and nonproductive areas.

2.2. Regional Studies

The most common method of mapping irrigated areas at regional scales is to use medium to coarse spatial resolution data from many time periods in the form of time series analysis. The image-based monitoring in the time domain relies heavily on the monitoring of crop evolution during the growing season by means of original spectral bands or vegetation indices. This requires knowledge of the growth schedules of crop types common to the area.

One of the most cost effective ways of monitoring irrigation in large areas is to use freely available vegetation index data from coarse resolution sensors like AVHRR and MODIS [19,21,47,48,49,50,51,52]. There are various ways to translate multi-temporal information to a map of irrigation. For example, Xiao *et al.* [48] developed a paddy rice mapping algorithm that uses time series of three vegetation indices, namely the Land Surface Water Index (LSWI), the Enhanced Vegetation Index (EVI) and NDVI, derived from MODIS images. This method relies on identification of the initial period of flooding and transplanting in paddy rice fields by exploiting the sensitivity of LSWI to the increased surface moisture during this period

It is also possible to use the results of multi-temporal vegetation indices directly in modern classification algorithms such as decision trees [20,49,51]. Classification accuracies in such applications range from 60 percent to better than 90 percent [20,21,48,51]. Notable regional differences in irrigated areas estimated this way stem from climate factors, localized precipitation patterns, and specific cropping practices [51]. Moreover, this form of classification benefits greatly from a multi-stage approach where the land cover is progressively classified [51] and from using ancillary information on climate, surface moisture status, elevation, natural land cover cartography, and crop area and water masks [20,21,53].

There are indications that multi-temporal image classification using the original spectral bands in addition to the band transformations leads to better classification results when satellite data with high temporal and spectral dimensionality such as MODIS are used [21,52]. For example, Thenkabail *et al.* [21] developed innovative approaches to obtain precise information on the onset, peak, senescence and duration of each irrigated and rainfed land use class. In particular, their results demonstrated that the value of the MODIS band 5 centered at 1,240 nm provided the best separability, followed by bands 2 (859 nm), 7 (2,130 nm), 6 (1,640 nm) and 3 (648 nm).

Algorithms that can automatically identify the location and the number of vegetation peaks in a time series plot improve classification accuracies. For example, Kamthonkiat *et al.* [53] developed a Peak Detector Algorithm to identify the number of vegetative peaks based on time lag between peak rainfall and peak NDVI. Their results indicate that in rainfed rice there is high correlation at a 40–50 day lag between peak rainfall and a single peak NDVI, while in irrigated rice there are multiple peaks and multiple correlations with low values for at least 90 days after peak rainfall.

One of the limitations of time series analysis of coarse spatial resolution data is the imprecise irrigated area estimates, especially in locations with small cultivated plots and fragmented landscapes. In these types of environments, remotely sensed estimates are generally higher than reported estimates because of: (a) inadequate accounting of irrigated areas in reported statistics, especially minor forms of irrigation; (b) definition issues involved in mapping using both remote sensing and statistics; (c) difficulties in arriving at precise estimates of irrigated area fractions; and (d) imagery resolution [19,20,21,52].

One approach to overcome these issues is to use high spatial resolution data such as Landsat to calibrate irrigated area from coarse resolution observations using regression [46,49]. Results from these calibration-based approaches to area correction suggest that while calibration seems to work well for homogeneous areas, for heterogeneous areas, where irrigated fields are scattered against a predominantly non-vegetated background, coarse resolution vegetation index values become rather low and consequently the regression relations are less reliable. However, comparison of multiple data sources improves confidence in the classification and identifies areas requiring more intensive fieldwork [49]. Another approach directly from the remotely sensed observations is to use a regression algorithm to derive the fraction of irrigated area within each pixel that had been identified as having irrigation [20]. However, both approaches establish the fact that both remote sensing and national statistical approaches require further refinement with a substantial investment of time and resources for ground-truthing.

2.3. Global Studies

In recent years, there have been several attempts to determine the spatial extent of irrigation at global scales. These attempts stem particularly from the crucial need for accurate estimates of irrigated areas for water use assessments and food security studies. One such attempt is the USGS Global Land Cover Map [22] that was generated based on 1km AVHRR observations between April 1992 and September1993. Among several land-cover legends that the database contains, the Global Ecosystems Legend identifies four irrigated classes: irrigated grasslands, rice paddies and fields, hot irrigated cropland and cool irrigated cropland. When combined, these classes provide one of the few sources of remotely-sensed information on spatial distribution at the global scale. The major shortcoming of the USGS map, however, is that irrigated areas were determined as part of a broader classification scheme. Thus the emphasis was primarily on other land-cover types while irrigated classes received less attention and thus lower classification accuracy. Vörösmarty and Sahagian [14] compared irrigated lands depicted by the USGS map to the country level reports of irrigated areas [54] and pointed to major uncertainties in classification and inventory due to the highly politicized nature of FAO data reports, as well as to the technical limitations of the more objective datasets made from remote sensing.

More recently, the European Space Agency (ESA) produced a moderate spatial resolution (300 m) global land cover product (GlobeCover) using Medium Resolution Imaging Spectrometer (MERIS) data acquired between mid-2005 and mid-2006 [55]. Several thematic classes within the GlobeCover legend identify irrigated crops, including flooded rice. The classification process includes unsupervised clustering of selected mosaics that are then temporally characterized based on computation of phenological parameters (start, end and duration of the vegetation period), using the time series of MERIS mosaics. In the last step, the classes were labeled with help from regional land cover experts, who advised on the interpretation of the spectral-temporal classes, and defined the improved labeling rules.

Thenkabail *et al.* [56] evaluated the utility of spectral matching techniques to determine historical land use and land cover (LULC) and irrigated area classes using the historical time series 0.1-degree AVHRR data without ground truth comparison. By using monthly continuous time series AVHRR data, they established the LULC for the period 1982–1985 and for the period 1996–1999. They used quantitative and qualitative spectral matching methods to identify and label LULC classes, including spectral similarity value, spectral correlation similarity, Euclidian distance, and modified spectral angle. The results show that spectral similarity was the best method, followed by spectral correlation. The other methods were more complex, and provided results that were either uncertain or no better. They used data from India's Central Board of Irrigation and Power for validation. The 1982–1985 irrigated area was 2,975,800 hectares, which was 8.5 percent higher than the non-remote-sensing-based area estimate by the Central Board for 1984 (2,743,638 hectares). These results helped establish the strengths of the spectral matching techniques in identifying and labeling LULC and irrigated area classes from the historical satellite sensor data for which little or no ground truth data are available. The spectral matching methods can be further strengthened by additional research involving a rich ground-based knowledge base. The study also implied that future application of this method at global scale is promising but may require a better potential irrigation index and the use of remotely-sensed skin temperature measurements.

Thenkabail *et al.* [57] also produced a Global Irrigated Area Map (GIAM) for the end of the last millennium. They used multiple satellite sensor and ancillary data including AVHRR 3 band and NDVI; SPOT VGT NDVI; monthly rainfall and temperature; elevation; Japanese Earth Resources Satellite-1 Synthetic Aperture Radar (JERS-1 SAR) data for the rain forests; and a map of global tree cover. In the first phase, the global land surfaces were segmented based on elevation, temperature and precipitation zones. Next, classification was performed within these segments using class spectra derived from unsupervised classification; these were matched with ideal or target spectra. Finally, the mixed classes were resolved based on decision tree algorithms and spatial modeling. Whenever that did not work, the problem class was used to mask and re-classify the original input data and the class identification and labeling protocol was repeated. The sub-pixel area calculations were performed by multiplying full-pixel areas with irrigated area fractions for every class.

The most important property of this methodology is that it delineates areas most likely to be rainfed. Such mapping is essential in the entire classification process as it can be overlaid with a map of green areas and only the non-rainfed green areas can then be considered to be irrigated. The methodology described here to create a map of Potentially Irrigated Areas (PIA) is based on Droogers and Aerts [10].

3. EMERGING PATTERNS FROM EXISTING STUDIES

Spatial resolution requirements

Irrigation is practiced in many forms around the world and the spatial resolution requirements go hand in hand with its heterogeneous distribution. Figure 1(A–C) shows ground level photographs of several landscapes from very heterogeneous small plot sizes in SE Asia to vast expanses of irrigated fields in the U.S. The purpose of these pictures is to illustrate the great variation in irrigation presence from subsistence farming to major enterprises. This diversity directly affects our ability to map irrigation with remote sensing. In general, to map irrigated croplands accurately, the spatial resolution of the sensor should be at or below the size of the fields of interest. However, enhanced information brought about by high spatial resolution data does not tell the whole story—automatic identification of irrigated lands also relies heavily on temporal availability of information. If the low spatial resolution option is chosen, then frequent observations are available, but the low resolving power of pixels would lead to the mixed pixel problem. Consider a single AVHRR 10 km pixel, which encompasses an area of 10,000 hectares, while a MODIS 500 meter pixel covers 25 hectares. Obviously in many cases only a fraction of a coarse resolution pixel falls in a particular land cover category, such as irrigation. We also know that spatial resolution of a sensor is supply driven, that is, spatial resolution is fixed because of sensor design and flight characteristics.

Research has shown that the finer the spatial resolution, the greater the accuracy of irrigated area class designations [7,58]. For example, a recent study of the irrigated areas of the Ogallala Aquifer in the United States based on Landsat imagery and 364 ground truth points has shown that 18.5 percent of the areas actually irrigated were not included (errors of omission) in the GIAM dataset at 10 km resolution but 20.4 percent of the non-irrigated areas were included as irrigated areas (errors of commission) [59]. The reason for the errors of omission was that at coarser resolutions, smaller, more fragmented patches of irrigated croplands could not be delineated. In contrast, the errors of commission resulted because the large pixel sizes of coarse resolutions can at times map patchy non-irrigated areas that surround irrigated areas as if they too were irrigated. In either case, the need for finer spatial resolution is essential to resolve the confusion.

Figure 1. Near ground photographs of irrigation practices in three different irrigated locations. (A) Irrigated rice paddies with a bore well in the background in north central India. The complex nature of this landscape with multiple land cover types surrounding irrigated fields illustrates the need for high spatial resolution data which may come at a cost of frequency of observations needed to map paddy fields. Image is provided courtesy of Eby Heller from McGill University. (B) A large irrigated alfalfa field in Utah (USA) and a rotating drum irrigation system. In these types of environments, the spatial requirements for irrigation mapping can be relaxed but the spectral properties of the image signal must be enhanced to distinguish irrigation. Photo courtesy of USDA NRCS. (C) A large irrigated landscape interspersed with multiple land cover types in Idaho (USA). This photograph illustrates the fact that even though many landscapes may be heterogeneous, the size of individual landscape elements (e.g., cultivated fields) determines the spatial resolution requirements of an irrigation mapping exercise. Photo courtesy of USDA NRCS.

To overcome the mixed pixel issue present in low spatial resolution data, several researchers have developed schemes to "unmix" land cover classes—including irrigation—from the remotely sensed signal to estimate the sub-pixel areas of irrigated lands. Two approaches are common in the literature to estimate these irrigated area fractions. The first is a two-stage approach and determines the subpixel composition of a class label following initial categorical labeling with a classification algorithm [7,58]. For example, Thenkabail *et al.* [58] developed innovative methods to compute sub-pixel areas from coarse resolution data. They used the global GIAM dataset at 10km resolution based on AVHRR data and the irrigated area map for India at 500 m based on MODIS data to test and verify their methods. The sub-pixel irrigated areas from coarse-resolution satellite sensor data were estimated by multiplying the full pixel irrigated areas with irrigated area fractions. The results indicated that the sub pixel areas provided significantly better correlation with the national statistics than full pixel areas, which were also shown to grossly overestimate reported areas.

The second approach attempts to derive the information on subpixel composition of irrigated pixels directly from the remotely sensed signal using a pixel decomposition method. For example, the authors in [20,62] reported on the use of land surface temperature data, especially the diurnal difference (day *vs.* night) in land surface temperature, and how this may relate to irrigated area intensity. As an example, Figure 2 shows that as the difference between minimum and maximum land surface temperatures decreases, irrigated crop intensity increases due to increased modulation of radiative temperature by vegetation/moisture presence. This information can be used to directly map the intensity of irrigated areas using land surface temperature sensors at high observation frequencies.

The spatial resolution dependent changes in irrigated area were also explored by Velpuri *et al.* [63] who determined how irrigated areas change with resolution (or scale) of imagery in the Krishna River basin in India. The irrigated areas were mapped using satellite sensor data at four resolutions: (a) AVHRR (10,000 m); (b) MODIS (500 m); (c) MODIS (250 m), and (d) Landsat (30 m). The proportion of irrigated areas relative to Landsat derived irrigation figures (9.36 million hectares for the Krishna basin) were (a) 95 percent using MODIS 250 meter data, (b) 93 percent using MODIS 500 meter data, and (c) 86 percent using AVHRR at 10,000 meters. The study also found that the precise locations of the irrigated areas were better established using finer spatial resolution data. This study suggested that "the finer the spatial resolution of the sensor, the greater was the irrigated area," since at finer spatial resolutions fragmented areas are better detected. Accuracies and errors were established consistently across the four resolutions mentioned above for three classes: surface water irrigation, ground water/conjunctive use irrigation, and no irrigation. The results showed that Landsat data provided significantly higher overall accuracy (84 percent) when compared to MODIS 500 m (77 percent), MODIS 250 m (79 percent), and AVHRR 10,000 m (63 percent).

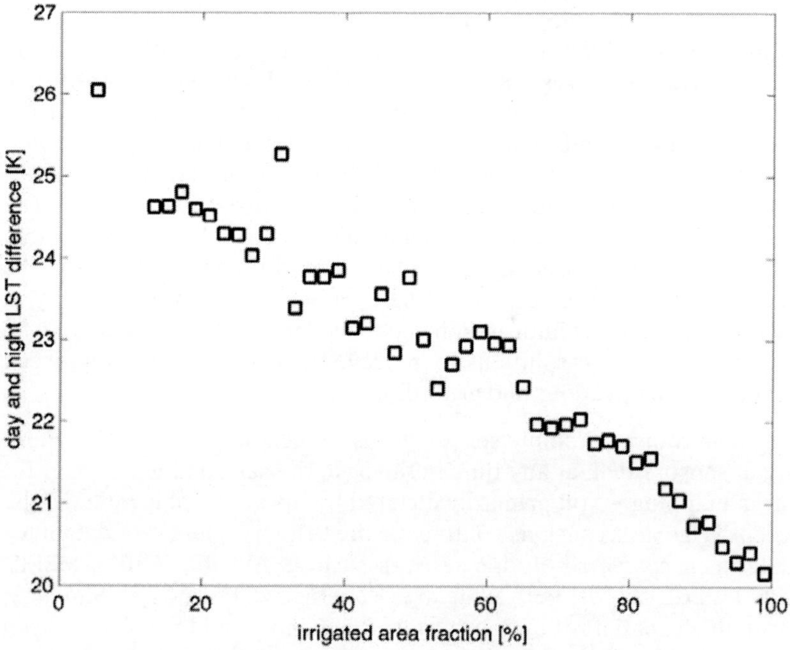

Figure 2. The relationship between diurnal difference (day minus night) in land surface temperature and subpixel intensity of irrigated area at a location in western US as observed by MODIS. As the irrigation intensity increases, the modulating effect of evaporation and associated cooling of the land surface becomes more pronounced. This indicates that the land surface temperature difference data which is routinely available from remote sensing could be used to map fractional coverage of irrigation using coarse resolution sensors. Ozdogan unpublished data.

Even if the satellite sensor has sufficient spatial resolution to map irrigated areas in a particular location, the pre-processing of the sensor data may distort the original resolution. For example, Tan *et al.* [64] showed that gridding artifacts between satellite observations and predefined grid cells strongly influence the local spatial properties of MODIS images. The sensor observation in any grid cell is only partially derived from the location of the cell, with the average overlap between observations and their grid cells being less than 30 percent. This mismatch between grid cells and observations has important implications for the use of reference data for the validation of MODIS products or the training of MODIS algorithms. When generating multi-date composites, gridding artifacts also introduce bias when spectral compositing criteria are used. The band-to-band registration for the different spatial resolutions of gridded MODIS data can be poor if the different resolutions of data are gridded before aggregation. In all cases it is imprecise to characterize the subpixel properties of the coarser resolution bands using the finer resolution bands due to poor correspondence in the areas from which the observations are derived. All of the band-to-band registration problems are minimized when the MODIS data are aggregated to coarser resolutions. When validating algorithm accuracy, data on the observation dimensions and the offsets between the grid cell and the observation should be included to ensure the quality of validation results. If this information is not available, MODIS data should be aggregated to coarser resolutions to improve the correspondence between the location of observations and grid cells.

Although the choice of remote sensing measurements today is from a wider variety of sensors than at any time in the past, the selection of a sensor for irrigation mapping is still primarily dictated by the geographic range of the study. For large areas such as continents, the primary sources of data have typically been coarse resolution sensors such as AVHRR, MODIS, MERIS, and SPOT VEGETATION with wide area coverage [24,65,66,67]. Similarly, data from high spatial resolution sensors like Landsat and SPOT have been used in support of local and regional scale applications that require increased spatial detail. What is less known, however, is the magnitude of errors incurred in irrigated area estimates as a function of the spatial resolutions of different sensors. Previous studies show that while coarsening the spatial resolution clearly leads to a loss of spatial detail [68], the magnitude of errors in area estimation as pixels increase in size is dependent on the spatial structure of the landscape [69,70].

Spectral data requirements

Remote sensing of irrigated lands over different geographic regions involves significant challenges, first in selecting spectral bands or indices that contain the maximum amount of irrigation-related information, then in relating this information to complex forms of irrigation presence. For instance, while satellite-derived indices are extremely useful in detecting vegetated areas in agricultural lands, the cause of spatial and temporal variability in biomass, whether as a result of rainfall or irrigation, is much less straightforward, hence more difficult to determine. A further complication may arise when only supplemental irrigation is practiced [71]. While, the characteristic shape of the fields where center pivot irrigation is practiced is obvious, this visual discrimination is neither evident nor can it easily be incorporated in automated image classification.

To overcome these challenges the characteristic attributes of irrigated lands, especially those characteristics that are observable with remotely sensed measurements must first be identified. One such attribute is related to vegetation "greenness." There is an overwhelming consensus that the NDVI is an important tool to monitor vegetation, including irrigated lands [65,72,73,74,75]. Studies of agricultural lands have shown that in many semi-arid regions, NDVI alone is a good indicator of irrigation [30,33,36,37], as well as irrigation status [35], crop condition [76] and large-scale vegetation dynamics [77].

While the NDVI signal associated with irrigation allows identification of irrigated lands, particularly in semi-arid areas characterized by a single irrigation period and simple land cover types, in many areas of the world there are multiple irrigation periods and several crop types with different schedules. In these locations, the NDVI signal associated with irrigation might be more subtle. Frequent observations of the NDVI from sensors like MODIS may help to map these lands and thus determine acquisition times for high spatial resolution imagery. However, difficult cases for distinguishing irrigated from non-irrigated crops occur in locations where the same crop type is grown with and without irrigation during the same growing season. Temporal NDVI profiles of both irrigated and non-irrigated crops in these locations may exhibit an identical pattern. While irrigated fields often exhibit greater greenness due to the constant availability of moisture, the NDVI difference between irrigated and non-irrigated fields is small and probably useless in distinguishing them. Thus, a more sensitive index may be required to make this distinction.

A large body of research of spectral remote sensing of vegetation canopies indicates that moisture stress in vegetation is strongly manifested in spectral indices related to chlorophyll content [78,79]. One such index, suggested by Gitelson et al. [80] to be used with the MODIS sensor, is the Green Index (GI) defined as:

$$GI = \rho_{nir} / \rho_{green} \tag{1}$$

where ρ_{green} is the reflectance in the green spectral region. The theoretical foundations of the GI are given in Gitelson *et al.*[78]. Briefly, it is based on evidence that in the green spectrum (centered around 510 nm) the specific absorption coefficient of chlorophylls is very high while green leaves are absorbing more than 80 percent (e.g., [81]). In contrast, depth of light penetration into leaves in the blue and red spectral ranges is one-fourth to one-sixth as much (e.g., [82]). Therefore, in the green spectrum, absorption of light is great enough to provide high sensitivity of the GI to chlorophyll content but much lower than in the blue and red to avoid light saturation [78].

The theoretical foundation for using the GI for irrigation assessment is that irrigated crops with very little or no soil moisture stress will exhibit higher chlorophyll content than non-irrigated crops that potentially experience moisture stress. To test this hypothesis and the sensitivity of the GI to the presence of irrigation (moisture), Ozdogan and Gutman [20] compared four vegetation indices in the midwestern US: NDVI, Enhanced Vegetation Index (EVI) [81], Wide Dynamic Range Vegetation Index (WDRVI) [82], and the GI, using the following Relative Sensitivity Index (RSI):

$$RSI = \frac{I_{irr} - I_{non}}{I_{non}(\max) - I_{non}(\min)} \tag{2}$$

In Equation (2), I_{irr} and I_{non} are the irrigated and non-irrigated values for each index, I, at each time period, normalized by the seasonal amplitude (maximum–minimum) of non-irrigated values of each index. RSI represents the difference between irrigated and non-irrigated index value compared to the seasonal maximum change non-irrigated value of the same index. Comparison of relative sensitivity of all four indices reveals that the GI shows the greatest sensitivity to irrigation presence during peak crop growth (Figure 3). Similar results were obtained by Gitelson *et al.* [83] in the same location. Recognizing the increased sensitivity to irrigation, we suggest that future mapping studies include the GI in the classification process.

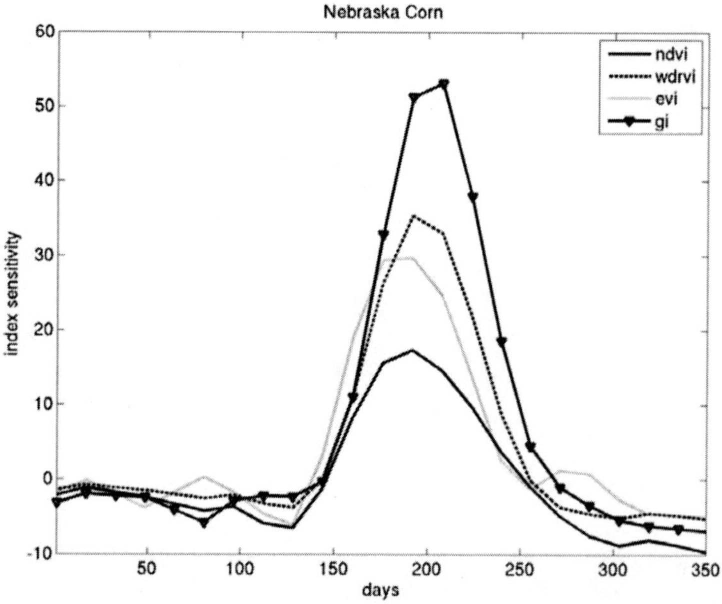

Figure 3. Sensitivity of four vegetation indices, namely NDVI, EVI, GI, and WDRVI, to irrigation presence as measured by Relative Sensitivity Index (RSI) in Nebraska for 2002. Each index was generated from an average response using a sample of approximately five individual sites. Please see the text for details on the RSI. Adapted from Ozdogan and Gutman [20].

As noted, irrigated areas are very dynamic in time and space and thus their spectral signatures quickly change across the growing seasons. Thenkabail *et al.* [21] exploited this feature of irrigated areas in a two-dimensional near infrared versus red band spectral reflectivity plots of different land cover types including that of irrigation (Figure 4). Typically, classes that have high near-infrared reflectivity and low red reflectivity are forests, agricultural lands and natural vegetation. Classes that occupy bright areas with high near-infrared and high red reflectivity are likely to be open/barren areas, sparse vegetation, dry vegetation, clouds and built-up areas. Classes that occupy wet areas have low near-infrared and low red reflectivity and are likely to be wetlands, moist lands, water bodies, cloud shadows and swamp forests. From irrigation perspective, these two dimensional spectral time curves provide very useful information on class behavior. For example, irrigated areas occupy the largest territory and the two dimensional plots provide very good indications of class temporal and spectral characteristics.

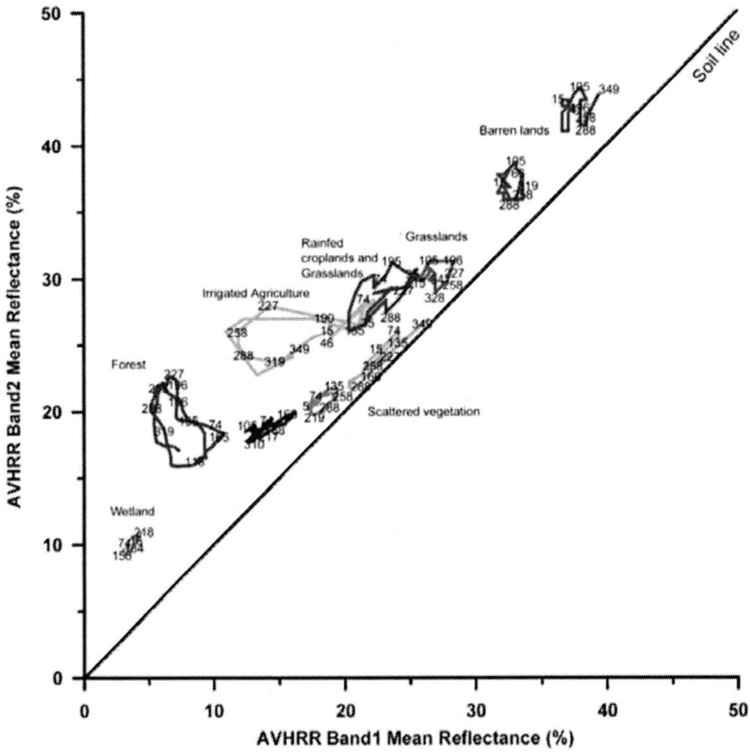

Figure 4. Space-time spiral curves (ST-SCs) in class identification and labeling. The ST-SCs track changes of time series over time and across space. The numbers seen in each class represent Julian date and each class moves around a 'territory' in 2D feature space over time. Adapted from Thenkabail *et al.* [57].

Another approach to spectral identification of irrigation is to compare each pixel's greenness value to its regional mean as measured by an index. Here region is loosely defined and can include a command area or an agro-ecological zone, and applies only to the same land cover type (e.g., agriculture). The assumption is that the pixels representing irrigated fields will have greater greenness than their non-irrigated counterparts within any region.

Although it's examples are rare and far in-between in irrigation mapping studies, there is also significant value in considering hyperspectral narrow band indices for further improving our understanding and accurate identification of irrigated areas. Also known as imaging spectroscopy, hyperspectral imaging has been used under laboratory conditions for over

100 years for identification of materials and their composition. Spectroscopy can be used to detect individual absorption features due to specific chemical bonds in a solid, liquid, gas including vegetation. The key to a successful use of narrow band spectrometers in identifying irrigation status of vegetation canopies will be through detection of moisture stress (which is often non-existent in irrigated areas) and through leaf chemical properties that are unique under artifical watering conditions.

Temporal data requirements

The mapping of irrigated lands with remote sensing is strongly affected by the timing of image acquisition and the number of images used [84]. The image-based monitoring relies heavily on the temporal monitoring of crop evolution during the growing season by means of vegetation indices. Thus, precise timing of image capture is necessary to effectively distinguish irrigated crops from each other and from other land cover types [37]. This requires knowledge of the growth schedules of crop types common to the area. For example, Pax-Lenney *et al.* [35] concluded that in addition to the number of images needed, timing of those images within the agrarian calendar proved to be more important than expected for lands under continuous cultivation in the Nile delta. In that study, nonproductive lands could not be identified with adequate accuracy by using images from low-growth seasons and their areas were overestimated by at least 400 percent with data sets consisting of two, three, or four low-growth-season images. Thus, even large multi-temporal datasets do not entirely mitigate the effects of timing interactions between images. Timing and image combinations restrict the accuracy of the identification of irrigated lands even with large image data sets. In some locations this is unfortunate, because low vegetation imagery is often the only cloud-free imagery available.

Most studies at local scales show that irrigated area overestimation problems decrease as more peak-season images are incorporated into the data sets. However, given data sets of equal numbers of images, peak-growth imagery may not always perform as well as combinations of peak and low growth-season imagery. The result is that in some cases, incorporating the variability or range of the NDVI values across seasons delineates irrigated from non-irrigated lands more accurately than maximum NDVI data alone can do [35].

These studies also suggest that the number of images needed to achieve a predetermined level of accuracy will differ between geographic regions. This emphasizes the fact that the number of images needed to monitor irrigated lands will vary by location, regional agricultural practices, and the feature of interest.

As stated, the quality of irrigation maps derived from remote sensing partly depends on the number and timing of images used. While there is a need for multi-temporal imagery in distinguishing summer crops, this analysis shows that in the study area, methods such as NDVI thresholding based on single images produces useful results, especially if guided by high frequency remote sensing observations like MODIS. This is particularly important for incorporating remote sensing technology into the future irrigation management practices in many of the world's semi arid areas. While remote detection of summer irrigated lands by this means may be simplistic, it provides an excellent example of the practical utility of remote sensing for summer irrigation monitoring.

4. AREAS THAT REQUIRE FURTHER RESEARCH

Use of Radar data

To form the basis for developing sustainable water management practices requires knowledge of the current extent, area, and the status of irrigated agricultural landscapes. Advances in remote sensing based mapping technologies have partially contributed in documenting and monitoring irrigated land use types. There are, however, several unresolved problems associated with mapping irrigated areas at regional to continental scales in regions with frequent cloud cover. As described in previous sections, optical remote sensing has been successfully used for classification of irrigated areas from local to global scales using visual interpretation and classification as their primary approach for extracting thematic information.

During the past decade, several radar sensors have been deployed in space. Though none were designed specifically for irrigation mapping, several investigations have demonstrated that the data provide unique information about the characteristics of irrigated landscapes [85,86,87]. First, the radar data can be acquired as frequently as possible without atmospheric interference and solar angle variations. Second, depending on the wavelength, the radar backscatter signal carries information about the moisture status of vegetated landscapes. For example, [85] discovered that the radar backscatter coefficient of rice fields, appears to have a significant temporal variation and that this variation can be used to identfy paddy rice fields. Similarly, [86] report a strong relationship between L-band backscatter and plant growth in manually planted rice areas although the radar response is strongly dependent on parameters related to the spatial distribution of the plants. Despite these artifacts, however, a correlation between plant growth and backscatter has been observed for all fields, independent of plant spacing and field orientation [86]. While these investigations established the physical foundation for understanding the relationship between radar backscatter in paddy rice fields, another study [87] suggests that rice production estimates require at least three radar acquisitions taken at three crop growth and development stages. However, if multi-parameter radar data are available, only two acquisitions may be sufficient.

Note that the combination of radar data together with optical data has not yet been exploited to its fullest extent. Radar is useful because of its sensitivity to soil moisture status, even in complex environments. Furthermore, radar data can be collected in almost all weather conditions, a characteristic that is especially important in areas with frequent cloud cover.

Use of passive microwaves for moisture status

Irrigated areas, by definition, contain increased soil moisture due to artificial watering of fields. Microwave remote sensing can be used to estimate soil moisture and, by extension, irrigation, on the basis of the large contrast that exists between the dielectric constant values for dry and wet soils [88,89]. Temporal monitoring of water availability at the soil root zone during growth periods could provide a way to pre-determine irrigated and non-irrigated fields. This information can then be refined using optical or radar sensors at increased spatial resolutions. Because microwaves penetrate clouds, microwave remote sensing also presents a good alternative to optical remote sensing for monitoring irrigation distribution, as well as the health and water needs of irrigated areas in humid locations with frequent cloud cover. One limitation of passive microwave data is their coarse spatial resolution (on the order of 25 km on a side). Nevertheless, these data can be used to pre-determine the locations of irrigated areas in a broader context using remote sensing.

Classification algorithms with many-to-one mapping capabilities

Because of the number of images to be processed and the need for multi-temporal data, the use of more automated methods based on supervised classification using machine learning algorithms is important. Neural networks (specifically Fuzzy ARTMAP: [90,91,92]), decision trees [24], and support vector machines (SVM) [93] are particularly powerful in handling information from multiple sources or dates to generate reliable classification results. One of the strengths of these machine-learning algorithms is the "many-to-one" mapping capability, meaning that any given output class can have many spectral manifestations, each of which is preserved internally within the trained set. For irrigation mapping, this is particularly useful, as most studies use only a limited number of classes and the non-irrigated class could encompass multiple land cover classes. The idea of many spectral manifestations can also be extended into the temporal domain, where irrigated productive agricultural fields and barren non-irrigated fields each with particular multi-date NDVI trajectories can be fed into a training set as individual training samples to classify the amount of irrigated land per study period.

Object-oriented classification

When using high spatial resolution data to map irrigated agricultural lands, traditional spectral based methods may result in rather poor or incorrect classification. This is because in this type of data, much information is contained in spatial relations of pixels and not in their spectral response or temporal change. Therefore, methods that exploit the relationship between a group of pixels in an attempt to generate improved classification are promising when analyzing high resolution data in fragmented environments. This form of classification is called the object-oriented approach. Unfortunately, there are no examples of application of object-oriented classification methods to map irrigated landscapes in the literature, although a few examples exist for mapping land cover types in cultivated settings.

In an object-oriented classification process, image objects (e.g., agricultural fields) can be created by means of automatic segmentation algorithms, or by ancillary sources of information such as cadastral cartography. In general, a segmentation algorithm would produce a space division conditioned by sensor attributes instead of geographic characteristics. Image classification by parcels has been widely used for agricultural applications [94,95]. Object-based image analysis can be accomplished following two different approaches: prior to classification (per-parcel) and after per-pixel classification. In the first approach, descriptive features of each parcel are first calculated and then parcels are classified as a whole using any of the available classification algorithms. In the latter approach, the frequency of the classified pixels within each parcel using either the majority pixel class as the assigned pixel label or a mixed label describing the heterogeneity of the parcel assigned [94,96].

Data Fusion

Information about the agricultural areas is currently being captured by a number of satellite-based sensors of different spatial, spectral, temporal, and radiometric characteristics. It would be beneficial to utilize information from multiple sources to more effectively map irrigated areas and the concept of merging data for the same scene from different sensors is called data fusion. The motivation behind data fusion is to generate an interpretation of the scene not obtainable with data from a single sensor, or to reduce the uncertainty associated with the data from individual sensors [98]. For example, for an image segmentation or classification task, the goal of fusing data from different sensors is to reduce the classification error rate obtained by single source classification.

Pohl and Van Genderen [98] suggest that before being able to implement and use an image fusion approach some critical questions need to be answered by the user including: What is the objective/application of the user? Which types of data are the most useful for meeting these needs? Which is the `best' technique of fusing these data types for that particular application? What are the necessary pre-processing steps involved? Which combination of the data is the most successful? These and other questions comprise a large number of parameters to be considered.

While there are not many examples of data fusion for classification problems in irrigated agricultural settings, a small review of the existing methodologies here is useful. In general, data fusion techniques can be grouped into two classes: (i) Colour related techniques, and (ii) Statistical/numerical methods [98]. The first comprises the color composition of three image channels in the RGB color space as well as more sophisticated color transformations. Statistical approaches, on the other hand, are developed on the basis of channel statistics including correlation and filters. Techniques like PCA and regression belong to this group. The numerical methods follow arithmetic operations such as image differencing and ratios but also adding of a channel to other image bands. A form of sophisticated numerical approach uses wavelets in a multiresolution environment. For example, [99] uses this wavelet approach to merge Radar data with information from optical sensors to improve the classification accuracies. The premise here is that while each of these data types are not sufficient in identifying irrigated areas alone, their joint information is powerful and improve classification accuracies [98,99].

In addition to these forms of traditional image fusion methods, there is an emerging trend in the literature of what would be called "information merging" from different satellite sources. The main difference here is the goal of image or information fusion is not to generate pixels with improved information content from multiple data sources but rather to use information from multiple sources to make informed decisions. Some of the best examples of this new form of data fusion are provided by [7,21,56]. For example, Ozdogan *et al.* [7] used information from coarse resolution high temporal frequency observations as a guide to determining the ideal acquisition timing of high spatial resolution data for mapping purposes. While irrigated area mapping was accomplished solely based on high resolution data, the fact that an additional sensor data with different capabilities was used to steer the acquisition times is sufficient to consider this application a form of data fusion. Another great example is provided by [21] who used temporal profiles from one set of observations to interpret land cover type in another sensor data. This form of information fusion is important in irrigated area mapping since the phenology (temporal profiles) of crops are crucial to their identification and mapping—not only from other crops, but also from other vegetation.

As noted, single date imagery might suffice (if acquired at right time of crop phenology) to map irrigated areas at local scales (e.g., corn *vs.* soybeans during critical growth phases in U.S. mid-west) but when the extent of mapping in regional/continental/global this will be highly inadequate. Here is where data fusion, along with non-optical data from microwave remote sensing and numerous secondary data, included in data fusion, will be invaluable.

5. CONCLUSIONS

Satellite data offer tremendous advantages for irrigated area mapping problems at various temporal and spatial scales. However, for more effective use of remote sensing, the analyst should be aware of the limitations and advantages of satellite data and should choose from the available irrigation mapping options accordingly. For example, methods that work in local areas may not lend themselves easily to regional and global applications. Although satellite remote sensing cannot provide the detailed information available from aerial photography or field studies, it can provide complimentary information to these conventional mapping techniques. It can also identify areas where changes are occurring and where more detailed information must be gathered.

Techniques for improving the identification of irrigated areas using remote sensing data include the use of multi-temporal imagery and ancillary data and these methods hold true across all spatial scales considered here. In general, it was found that multi-temporal imagery and ancillary information such as climate, soils, or slope improved irrigation classification. Multi-temporal imagery provides the greatest accuracy for delineating irrigation from other land cover types. While the ideal dates will differ depending on the type and location of irrigation being studied, it is possible to make use of high frequency observations at coarse spatial resolutions even in local area investigations. In general, ancillary data, when available, improves irrigation classification. Machine learning and rule-based classification methods generally provide better results than conventional statistical classification approaches. If detailed information is required at a finer spatial resolution than the satellite sensor can provide, then a subpixel classification scheme should be used. Several methods have been developed for this purpose. In conclusion, mapping of irrigated areas with satellite imagery is difficult but possible. As with many remote sensing problems, satisfactory results may require techniques specific to the location on a case by case basis.

Finally, multisensor data fusion provides an effective paradigm for remote sensing applications by synthesizing data from multiple sensors or sources. For example, high-resolution Landsat style observations can be integrated with low-resolution high temporal frequency observations to complement each other for the improved information extraction in irrigated settings. High spatial resolution images can also be processed by the state-of-the-art image segmentation algorithms to generate individual objects that often correspond to physically meaningful entities, e.g., a cropland unit growing wheat.

ACKNOWLEDGEMENTS

This research was partly funded by the NASA Applications Program Grant NNX08AM69G, awarded to Mutlu Ozdogan. The early comments of Annemarie Schneider greatly improved this manuscript. Mutlu Ozdogan is also indebted to the guest editor Prasad Thenkabail for encouraging the writing of this review and including it in this special issue. Finally, Elvis Wang at the MDPI editorial office is greatly acknowledged for his meticulous editing that improved the readability of this document.

REFERENCES AND NOTES

1. Gleick, P.H. Global freshwater resources: Soft-path solutions for the 21st century. *Science* **2003**, *302*, 1524.
2. Rosegrant, M.W.; Meijer, S.; Cline, S.A. *International Model for Policy Analysis of Agricultural Commodities and Trade (IMPACT): Model Description*; IFPRI: Washington, DC, USA, 2002.
3. Cai, X.; Rosegrant, M.W. Global water demand and supply projections. *Water Int.* **2002**, *27*, 159–169.
4. Stockholm International Water Institute. Available online: http://www.siwi.org/ (accessed on 17 June 2010).

5. Boucher, O.; Myhre, G.; Myhre, A. Direct human influence of irrigation on atmospheric water vapour and climate.*Climate Dynamics* **2004**, *22*, 597–603.
6. Gordon, L.J.; Steffen, W.; Jönsson, B.F.; Folke, C.; Falkenmark, M.; Johannessen, A. Human modification of global water vapor flows from the land surface. *Proc. Nat. Acad. Sci. USA* **2005**, *102*, 7612.
7. Ozdogan, M.; Woodcock, C.E.; Salvucci, G.D.; Demir, H. Changes in summer irrigated crop area and water use in southeastern Turkey from 1993 to 2002: Implications for current and future water resources. *Water Resour. Manag.***2006**, *20*, 467–488.
8. Wisser, D.; Frolking, S.; Douglas, E.M.; Fekete, B.M.; Vörösmarty, C.J.; Schumann, A.H. Global irrigation water demand: Variability and uncertainties arising from agricultural and climate data sets. *Geophys. Res. Lett.* **2008**, *35*, L24408.
9. Alcamo, J.; Döll, P.; Henrichs, T.; Kaspar, F.; Lehner, B.; Rösch, T.; Siebert, S. Global estimates of water withdrawals and availability under current and future "business-as-usual" conditions. *Hydrol. Sci. J.* **2003**, *48*, 339–348.
10. Droogers, P.; Aerts, J. Adaptation strategies to climate change and climate variability: A comparative study between seven contrasting river basins. *Phys. Chem. Earth, Parts A/B/C* **2005**, *30*, 339–346.
11. Jones, R.N. Analysing the risk of climate change using an irrigation demand model. *Climate Res.* **2000**, *14*, 89–100.
12. Rosenzweig, C.; Strzepek, K.M.; Major, D.C.; Iglesias, A.; Yates, D.N.; McCluskey, A.; Hillel, D. Water resources for agriculture in a changing climate: International case studies. *Glob. Environ. Change Part A* **2004**, *14*, 345–360.
13. Vörösmarty, C.J.; Green, P.; Salisbury, J.; Lammers, R.B. Global water resources: Vulnerability from climate change and population growth. *Science* **2000**, *289*, 284–288.
14. Vörösmarty, C.J.; Sahagian, D. Anthropogenic disturbance of the terrestrial water cycle. *BioScience* **2000**, *50*, 753–765.
15. Moore, N.; Rojstaczer, S. Irrigation's influence on precipitation—Texas High Plains, U.S.A. *Geophys. Res. Lett.* **2002**, *29*.
16. Kueppers, L.M.; Snyder, M.A.; Sloan, L.C. Irrigation cooling effect: Regional climate forcing by land-use change.*Geophys. Res. Lett.* **2007**, *34*, L03703.

17. Vörösmarty, C.J. Global water assessment and potential contributions from Earth Systems Science. *Aqua. Sci.-Res. Across Boundaries* **2002**, *64*, 328–351.
18. Bastiaanssen, W.G.M.; Molden, D.J.; Makin, I.W. Remote sensing for irrigated agriculture: Examples from research and possible applications. *Agr. Water Manag.* **2000**, *46*, 137–155.
19. Thenkabail, P.S.; Dheeravath, V.; Biradar, C.M.; Gangalakunta, O.R.P.; Noojipady, P.; Gurappa, C.; Velpuri, M.; Gumma, M.; Li, Y. Irrigated area maps and statistics of India using remote sensing and national statistics. *Remote Sensing* **2009**, *1*, 50–67.
20. Ozdogan, M.; Gutman, G. A new methodology to map irrigated areas using multi-temporal MODIS and ancillary data: An application example in the continental US. *Remote Sens. Environ.* **2008**, *112*, 3520–3537.
21. Thenkabail, P.S.; Schull, M.; Turral, H. Ganges and Indus river basin land use/land cover (LULC) and irrigated area mapping using continuous streams of MODIS data. *Remote Sens. Environ.* **2005**, *95*, 317–341.
22. Loveland, T.R.; Reed, B.C.; Brown, J.F.; Ohlen, D.O.; Zhu, Z.; Yang, L.; Merchant, J.W. Development of a global land cover characteristics database and IGBP DISCover from 1 km AVHRR data. *Int. J. Remote Sens.* **2000**, *21*, 1303–1330.
23. Pittman, K.; Hansen, M.C.; Becker-Reshef, I.; Potapov, P.V.; Justice, C.O. Estimating global cropland extent with multi-year MODIS data. *Remote Sensing* **2010**, *2*, 1844–1863.
24. Friedl, M.A.; McIver, D.K.; Hodges, J.C.F.; Zhang, X.Y.; Muchoney, D.; Strahler, A.H.; Woodcock, C.E.; Gopal, S.; Schneider, A.; Cooper, A. Global land cover mapping from MODIS: Algorithms and early results. *Remote Sens. Environ.* **2002**, *83*, 287–302.
25. FAO. Topics: Irrigation. Water Development and Management Unit, United Nations Food and Agricultural Organization. Available online: http://www.fao.org/nr/water/topics_irrigation.html (acces sed on 17 June 2010).
26. Huston, D.M.; Titus, S.J. *An Inventory of Irrigated Lands for Selected Counties within the State of California Based on Landsat and Supporting Aircraft Data*; Space Sciences Laboratory Semi-Annual Progress Report; Space Sciences Laboratory Series 16; University of California: Berkeley, CA, USA, 1975; p. 23.

27. Draeger, W.U. *Monitoring Irrigated Land Acreage Using Landsat Imagery: An Application Example*; USGS Open-file Report No. 76-630; USGS: Sioux Falls, SD, USA, 1976; p. 23.
28. Heller, R.C.; Johnson, K.A. Estimating irrigated land acreage from Landsat imagery. *Photogramm. Eng. Remote Sensing* **1979**, *45*, 1379–1386.
29. Thiruvengadachari, S. Satellite sensing of irrigation patterns in semiarid areas—An Indian study. *Photogramm. Eng. Remote Sensing* **1981**, *47*, 1493–1499.
30. Kolm, K.E.; Case, H.L. The identification of irrigated crop types and estimation of acreages from Landsat imagery. *Photogramm. Eng. Remote Sensing* **1984**, *50*, 1479–1490.
31. Thelin, G.P.; Heimes, F.J. *Mapping Irrigated Cropland from Landsat Data for Determination of Water Use from the High Plains Aquifer in Parts of Colorado, Kansas, Nebraska, New Mexico, Oklahoma, South Dakota, Texas, and Wyoming*; US Geological Survey Professional Paper; USGS: Washington, DC, USA, 1987.
32. Rundquist, D.; Hoffman, R.; Carlson, M.; Cook, A. The Nebraska center-pivot inventory—An example of operational satellite remote sensing on a long term basis. *Photogramm. Eng. Remote Sensing* **1989**, *55*, 587–590.
33. Eckhardt, D.W.; Verdin, J.P.; Lyford, G.R. Automated update of an irrigated lands GIS using SPOT HRV imagery. *Photogramm. Eng. Remote Sensing* **1990**, *56*, 1515–1522.
34. Ram, B.; Kolarkar, A.S. Remote sensing application in monitoring land-use changes in arid Rajasthan. *Int. J. Remote Sens.* **1993**, *14*, 3191–3200.
35. Pax-Lenney, M.; Woodcock, C.E.; Collins, J.B.; Hamdi, H. The status of agricultural lands in Egypt: The use of multitemporal NDVI features derived from Landsat TM. *Remote Sens. Environ.* **1996**, *56*, 8–20.
36. Abuzar, M.; McAllister, A.; Morris, M. Classification of seasonal images for monitoring irrigated crops in a salinity-affected area of Australia. *Int. J. Remote Sens.* **2001**, *22*, 717–726.
37. Beltran, C.M.; Belmonte, A.C. Irrigated crop area estimation using Landsat TM imagery in La Mancha, Spain. *Photogramm. Eng. Remote Sensing* **2001**, *67*, 1177–1184.

38. Keene, K.M.; Conley, C.D. Measurement of irrigated acreage in western Kansas from Landsat images. *Environ. Geol.* **1980**, *3*, 107–116.
39. Haack, B.; Wolf, J.; English, R. Remote sensing change detection of irrigated agriculture in Afghanistan. *Geocarto Int.* **1998**, *13*, 65–75.
40. Rao, P.P.N.; Mohankumar, A. Cropland inventory in the command area of Krishnarajasagar project using satellite data. *Int. J. Remote Sens.* **1994**, *15*, 1295–1305.
41. El-Magd, I.A.; Tanton, T.W. Improvements in land use mapping for irrigated agriculture from satellite sensor data using a multi-stage maximum likelihood classification. *Int. J. Remote Sens.* **2003**, *24*, 4197–4206.
42. Kauth, R.J.; Thomas, G.S. The tasseled cap: graphic description of the spectral-temporal development of agricultural crops as seen by Landsat. In Proceedings of the Symposium on Machine Processing of Remotely Sensed Data, West Lafayette, IN, USA, , June 29–July 1, 1976; pp. 41–51.
43. Manavalan, P.; Kesavasamy, K.; Adiga, S. Irrigated crops monitoring through seasons using digital change detection analysis of IRS-LISS 2 data. *Int. J. Remote Sens.* **1995**, *16*, 633–640.
44. Starbuck, M.J.; Tamayo, J. Monitoring vegetation change in Abu Dhabi Emirate from 1996 to 2000 and 2004 using Landsat satellite imagery. *Arab Gulf J. Scientific Res.* **2007**, *25*, 71–80.
45. Simonneaux, V.; Duchemin, B.; Helson, D.; Raki, E.; Olioso, A.; Chehbouni, A. The use of high-resolution image time series for crop classification and evapotranspiration estimate over an irrigated area in central Morocco. *Int. J. Remote Sens.* **2008**, *29*, 95–116.
46. Akbari, M.; Mamanpoush, A.; Gieske, A.; Miranzadeh, M.; Torabi, M.; Salemi, H.R. Crop and land cover classification in Iran using Landsat 7 imagery. *Int. J. Remote Sens.* **2006**, *27*, 4117–4135.
47. Toomanian, N.; Gieske, A.S.M.; Akbary, M. Irrigated area determination by NOAA-Landsat upscaling techniques, Zayandeh River Basin, Isfahan, Iran. *Int. J. Remote Sens.* **2004**, *25*, 4945–4960.
48. Xiao, X.; Boles, S.; Liu, J.; Zhuang, D.; Frolking, S.; Li, C.; Salas, W.; Moore Iii, B. Mapping paddy rice agriculture in southern China using multi-temporal MODIS images. *Remote Sens. Environ.* **2005**, *95*, 480–492.

49. Biggs, T.W.; Thenkabail, P.S.; Gumma, M.K.; Scott, C.A.; Parthasaradhi, G.R.; Turral, H.N. Irrigated area mapping in heterogeneous landscapes with MODIS time series, ground truth and census data, Krishna Basin, India. *Int. J. Remote Sens.* **2006**, *27*, 4245–4266.
50. Alexandridis, T.K.; Zalidis, G.C.; Silleos, N.G. Mapping irrigated area in Mediterranean basins using low cost satellite Earth Observation. *Comput. Electron. Agr.* **2008**, *64*, 93–103.
51. Wardlow, B.D.; Egbert, S.L. Large-area crop mapping using time-series MODIS 250 m NDVI data: An assessment for the US Central Great Plains. *Remote Sens. Environ.* **2008**, *112*, 1096–1116.
52. Dheeravath, V.; Thenkabail, P.S.; Chandrakantha, G.; Noojipady, P.; Reddy, G.P.O.; Biradar, C.M.; Gumma, M.K.; Velpuri, M. Irrigated areas of India derived using MODIS 500 m time series for the years 2001-2003. *ISPRS J. Photogramm. Remote Sens.* **2009**, *65*, 42–59.
53. Kamthonkiat, D.; Honda, K.; Turral, H.; Tripathi, N.K.; Wuwongse, V. Discrimination of irrigated and rainfed rice in a tropical agricultural system using SPOT VEGETATION NDVI and rainfall data. *Int. J. Remote Sens.* **2005**, *26*, 2527–2547.
54. FAO. FAOSTAT-PC, FAO Statistics on diskette. Food and Agriculture Organization of the United Nations: Rome, Italy, 1998.
55. Arino, O.; Gross, D.; Ranera, F.; Bourg, L.; Leroy, M.; Bicheron, P.; Latham, J.; Di Gregorio, A.; Brockman, C.; Witt, R. GlobCover: ESA service for global land cover from MERIS. In Proceedings of IEEE International Geoscience and Remote Sensing Symposium, Barcelona, Spain, 23–28 July 2007; pp. 2412–2415.
56. Thenkabail, P.S.; Parthasaradhi, G.; Biggs, T.W.; Gumma, M.K.; Turral, H. Spectral matching techniques to determine historical land use/land cover (LULC) and irrigated areas using time-series AVHRR Pathfinder datasets in the Krishna river basin, India. *Photogramm. Eng. Remote Sensing* **2007**, *73*, 1029–1040.
57. Thenkabail, P.S.; Biradar, C.M.; Noojipady, P.; Dheeravath, V.; Li, Y.J.; Velpuri, M.; Gumma, M.; Gangalakuntag, O.R.P.; Turral, H.; Cai, X.L.; Vithanage, J.; Schull, M.A.; Dutta, R. Global irrigated area map (GIAM), derived from remote sensing, for the end of the last millennium. *Int. J. Remote Sens.* **2009**, *30*, 3679–3733.
58. Thenkabail, P.S.; Biradar, C.M.; Noojipady, P.; Cai, X.; Dheeravath, V.; Li, Y.; Velpuri, M.; Gumma, M.K.; Pandey, S. Sub-pixel area

calculation methods for estimating irrigated areas. *Sensors* **2007**, *7*, 2519–2538.

59. Kurz, B.; Seelan, S.K. Use of remote sensing to map irrigated agriculture in areas overlying the Ogallala Aquifer, U.S. In *Remote Sensing of Global Croplands for Food Security*; Thenkabail, P.S., Turral, H., Lyon, J.G., Biradar, C., Eds.; CRC Press: Boca Raton, FL, USA, 2009.
60. Lobell, D.B.; Asner, G.P.; Ortiz-Monasterio, J.I.; Benning, T.L. Remote sensing of regional crop production in the Yaqui Valley, Mexico: Estimates and uncertainties. *Agr. Ecosyst. Environ.* **2003**, *94*, 205–220.
61. Ozdogan, M. Mapping crop types from temporal unmixing of MODIS data: The case for Independent Component Analysis. *Remote Sens. Environ.* **2010**, *114*, 1190–1204.
62. Hutmacher, R.B. Infrared thermometry for canopy temperature measurements: applications and limitation in irrigation scheduling. In *Sensor and Techniques for Irrigation Management*; California State University: Fresno, CA, USA, 1988; pp. 19–22.
63. Velpuri, N.M.; Thenkabail, P.S.; Gumma, M.K.; Biradar, C.M.; Dheeravath, V.; Noojipady, P.; Yuanjie, L. Influence of Resolution in Irrigated Area Mapping and Area Estimations. *Photogramm. Eng. Remote Sensing* **2009**, *75*, 1383–1396.
64. Tan, B.; Woodcock, C.E.; Hu, J.; Zhang, P.; Ozdogan, M.; Huang, D.; Yang, W.; Knyazikhin, Y.; Myneni, R.B. The impact of gridding artifacts on the local spatial properties of MODIS data: Implications for validation, compositing, and band-to-band registration across resolutions. *Remote Sens. Environ.* **2006**, *105*, 98–114.
65. De Fries, R.S.; Hansen, M.; Townshend, J.R.G.; Sohlberg, R. Global land cover classifications at 8 km spatial resolution: the use of training data derived from Landsat imagery in decision tree classifiers. *Int. J. Remote Sens.* **1998**,*19*, 3141–3168.
66. Roy, D.P.; Lewis, P.E.; Justice, C.O. Burned area mapping using multi-temporal moderate spatial resolution data—A bi-directional reflectance model-based expectation approach. *Remote Sens. Environ.* **2002**, *83*, 263–286.

67. Frolking, S.; Xiao, X.; Zhuang, Y.; Salas, W.; Li, C. Agricultural land-use in China: A comparison of area estimates from ground-based census and satellite-borne remote sensing. *Glob. Ecol. Biogeogr.* **1999**, *8*, 407–416.
68. Woodcock, C.E.; Strahler, A.H. The factor of scale in remote sensing. *Remote Sens. Environ.* **1987**, *21*, 311–332.
69. Moody, A.; Woodcock, C.E. The influence of scale and the spatial characteristics of landscapes on land-cover mapping using remote sensing. *Landscape Ecol.* **1995**, *10*, 363–379.
70. Turner, M.G. Landscape ecology: the effect of pattern on process. *Ann. Rev. Ecol. Systemat.* **1989**, *20*, 171–197.
71. Droogers, P. *Global Irrigated Area Mapping: Overview and Recommendations*; Working Paper 36; International Water Management Institute: Colombo, Sri Lanka, 2002.
72. Goward, S.N.; Markham, B.; Dye, D.G.; Dulaney, W.; Yang, J. Normalized difference vegetation index measurements from the Advanced Very High Resolution Radiometer. *Remote Sens. Environ.* **1991**, *35*, 257–277.
73. Justice, C.O.; Townshend, J.R.G.; Holben, B.N.; Tucker, C.J. Analysis of the phenology of global vegetation using meteorological satellite data. *Int. J. Remote Sens.* **1985**, *6*, 1271–1318.
74. Myneni, R.B.; Hall, F.G.; Sellers, P.J.; Marshak, A.L. The interpretation of spectral vegetation indexes. *IEEE Trans. Geosci. Remote Sens.* **1995**, *33*, 481–486.
75. Tucker, C.J. Red and photographic infrared linear combinations for monitoring vegetation. *Remote Sens. Environ.* **1979**,*8*, 127–150.
76. Wiegand, C.L.; Everitt, J.H.; Richardson, A.J. Comparison of multispectral video and SPOT-1 HRV observations for cotton affected by soil salinity. *Int. J. Remote Sens.* **1992**, *13*, 1511–1525.
77. Tucker, C.J.; Holben, B.N.; Goff, T.E. Intensive forest clearing in Rondonia, Brazil, as detected by satellite remote sensing. *Remote Sens. Environ.* **1984**, *15*, 255–261.
78. Gitelson, A.A. Relationships between leaf chlorophyll content and spectral reflectance and algorithms for non-destructive chlorophyll assessment in higher plant leaves. *J. Plant Physiol.* **2003**, *160*, 271–282.

79. Zarco-Tejada, P.J.; Miller, J.R.; Mohammed, G.H.; Noland, T.L.; Sampson, P.H. Vegetation stress detection through chlorophyll a+b estimation and fluorescence effects on hyperspectral imagery. *J. Environ. Quality* **2002**, *31*, 1433–1411.
80. Gitelson, A.A.; Vina, A.; Ciganda, V.; Rundquist, D.C.; Arkebauer, T.J. Remote estimation of canopy chlorophyll content in crops. *Geophys. Res. Lett* **2005**, *32*, L08403.
81. Gitelson, A.; Merzlyak, M.N. Quantitative estimation of chlorophyll-a using reflectance spectra: experiments with autumn chestnut and maple leaves. *J. Photochem. Photobiol. B: Biol.* **1994**, *22*, 247–252.
82. Merzlyak, M.N.; Gitelson, A. Why and what for the leaves are yellow in autumn? On the interpretation of optical spectra of senescing leaves (Acer platanoides L.). *J. Plant Physiol.* **1995**, *145*, 315–320.
83. Huete, A.; Justice, C.; Van Leeuwen, W. MODIS vegetation index (MOD13) algorithm theoretical basis document. NASA Goddard Space Flight Centre: Greenbelt, MD, USA, 1999.
84. Pax-Lenney, M.; Woodcock, C.E. The effect of spatial resolution on the ability to monitor the status of agricultural lands. *Remote Sens. Environ.* **1997**, *61*, 210–220.
85. Ribbes, F.; Toan, T.L. Rice field mapping and monitoring with RADARSAT data. *Int. J. Remote Sens.* **1999**, *20*, 745–765.
86. Rosenqvist, A. Temporal and spatial characteristics of irrigated rice in JERS-1 L-band SAR data. *Int. J. Remote Sens.* **1999**, *20*, 1567–1587.
87. Shao, Y.; Fan, X.; Liu, H.; Xiao, J.; Ross, S.; Brisco, B.; Brown, R.; Staples, G. Rice monitoring and production estimation using multitemporal RADARSAT. *Remote Sens. Environ.* **2001**, *76*, 310–325.
88. Njoku, E.G.; Wilson, W.J.; Yueh, S.H.; Dinardo, S.J.; Li, F.K.; Jackson, T.J.; Lakshmi, V.; Bolten, J. Observations of soil moisture using a passive and active low-frequency microwave airborne sensor during SGP99. *IEEE Trans. Geosci. Remote Sens.* **2002**, *40*, 2659–2673.
89. Lakhankar, T.; Krakauer, N.; Khanbilvardi, R. Applications of microwave remote sensing of soil moisture for agricultural applications. *Int. J. Terraspace Sci. Eng.* **2009**, *2*, 81–91.
90. Carpenter, G.A.; Gjaja, M.N.; Gopal, S.; Woodcock, C.E. ART neural networks for remote sensing: vegetation classificationfrom Landsat

TM and terrain data. *IEEE Trans. Geosci. Remote Sens.* **1997**, *35*, 308–325.

91. Gopal, S.; Woodcock, C.E.; Strahler, A.H. Fuzzy Neural Network Classification of Global Land Cover from a 1 AVHRR Data Set. *Remote Sens. Environ.* **1999**, *67*, 230–243.
92. Abuelgasin, A. Change detection using adaptive fuzzy neural networks—Environmental damage assessment after the Gulf War. *Remote Sens. Environ.* **1999**, *70*, 208–223.
93. Foody, G.M.; Mathur, A. A relative evaluation of multiclass image classification by support vector machines. *IEEE Trans. Geosci. Remote Sens.* **2004**, *42*, 1335–1343.
94. Berberoglu, S.; Lloyd, C.D.; Atkinson, P.M.; Curran, P.J. The integration of spectral and textural information using neural networks for land cover mapping in the Mediterranean. *Comput. Geosci.* **2000**, *26*, 385–396.
95. Walter, V. Object-based classification of remote sensing data for change detection. *ISPRS J. Photogramm. Remote Sens.* **2004**, *58*, 225–238.
96. Aplin, P.; Atkinson, P.M.; Curran, P.J. Fine spatial resolution satellite sensor imagery for land cover mapping in the United Kingdom. *Remote Sens. Environ.* **1999**, *68*, 206–216.
97. van Laake, P.; Skutsch, M.; McCall, M.K. *A Sourcebook of Methods and Procedures for Monitoring and Reporting Anthropogenic Greenhouse Gas Emissions and Removals Caused by Deforestation, Gains and Losses of Carbon Stocks in Forests Remaining Forests, and Forestation*; GOFC-GOLD Report Version COP15-1; GOFC-GOLD Project Office, Natural Resources Canada: Alberta, AB, Canada; p. 197.
98. Pohl, C.; Van Genderen, J.L. Multisensor image fusion in remote sensing: Concepts, methods and applications. *Int. J. Remote Sens.* **1998**, *19*, 823–854.
99. Gungor, O.; Shan, J. An optimal fusion approach for optical and SAR images. In Proceedings of the ISPRS Commission VII Mid-term Symposium "Remote Sensing: From Pixels to Processes", Enschede, The Netherlands, 8–11 May 2006; pp. 111–116.

9

CHAPTER

HARNESSING EARTH OBSERVATION AND SATELLITE INFORMATION FOR MONITORING DESERTIFICATION, DROUGHT AND AGRICULTURAL ACTIVITIES IN DEVELOPING COUNTRIES

Humberto Barbosa[1], Carolien Tote[2], Lakshmi Kumar[3] and Yazidhi Bamutaze[4]

[1] Universidade Federal de Alagoas (UFAL), LAPIS, Brazil
[2] Flemish Institute for Technological Research (VITO), Centre for Remote Sensing and Earth Observation, Boeretang, Mol, Belgium
[3] Atmospheric Science Research Laboratory, SRM University, India
[4] Department of Geography, Geo-Informatics and Climatic Sciences, Makerere University, Kampala, Uganda

1. INTRODUCTION

With the drastic advances in technology over the past decades, the availability, as well as the quantity, of large data sets for research in almost every scientific field has increased dramatically. More specifically, the availability of earth observation-based imagery data and satellite information for research purposes and practical applications has grown with many organizations such as the European Organisation for the Exploitation of Meteorological Satellites (EUMETSAT), the National Aeronautics Space Administration (NASA), the National Oceanic and Atmospheric administration (NOAA), the Flemish Institute for

Technological Research (VITO), etc. for example, through GEONETCast, which is part of the core Global Earth Observation System of Systems (GEOSS), the users do not need to repeatedly build ground receiving stations for different satellites [1]. However, despite a wealth of remotely sensed data provided by GEONETCast, investments in science technology and innovations is often a low priority for decision and policy makers in most developing countries in Africa, Asia and global emerging economies like India and Brazil. Yet, most of these countries face serious environmental risks and development challenges which require reliable and timely access to accurate Earth Observation (EO) data and derived environmental information for their sustainable development. In particular, there is a clear need for research on the integration and utility of remote sensing data and products into the risk assessment cycle, scenario development and impact forecasting, in view of global (climate) change [2].

Climate alterations, although global in nature, may have different impacts in different regions of the world. Reports of Intergovernmental Panel on Climate Change (IPCC) showed that the global average surface temperature has increased over the 20th century by about 0.6ºC [3]. Global Circulation Models have projected that this rise in temperature may increase to a range of 1.5 - 5.8ºC by the end of the century. Results of crop growth modeling under climate change scenarios suggest that agriculture, and thus human well-being, will be negatively affected by climate change, especially in developing countries [4]. The vulnerability to drought and land degradation has increased in the past decades and this is especially true due to increased population pressure and limited livelihood options in drought-prone areas [5]. Significant gaps in observing systems exist, especially in developing countries, and timely access to both surface-based and space observations is still a challenge in many locations.

Global-scale population growth and economic development will have a large impact on water supply and demand, and it is necessary to understand the interactions between climate change and variability, hydrology and human systems, in order to have a view on future water vulnerabilities [6]. Since developing countries will become more susceptible to climate variability and drought, it is essential to develop climate (impact) monitoring services. A climate service involves broad partnerships of producer and user organizations, climate scientists, climate service providers, economists and social scientists. It provides an opportunity to interlink global, national and regional information systems; to provide essential information to policy makers, decision takers and to the public in general at regional and local scales, and a provide for a distributed decision-relevant research and development capability [7]. The climate service for the developing countries might focus on collaborative problem solving. Also capacity building and the improvement of infrastructure, related to the acquisition of advanced remote sensing technologies and the installation of satellite receiving stations by, is needed.

The gist of this chapter is therefore to strengthen the capacities of the regional scientific community to provide stakeholders from drought to agricultural activities with satellite information that is directly useful to improve decision-making in the context of the developing countries. The chapter is based on four case studies from Africa, South America, India and Europe to demonstrate the utility of satellite imagery data obtained from free or low cost platforms in providing information to address the above environmental issues, which are critical particularly in developing countries. Based on the expertise, experience and interest of a regional network of scientists from Uganda, Brazil, India, and Europe, the chapter focuses on harnessing satellite remote sensing resources and products for drought monitoring of areas subject to or in risk of land degradation processes, agricultural productivity and drought assessment. The following paragraphs include a series of example cases where time series of satellite imagery is used to monitor the impact of climate (change):

i. Degradation monitoring over South America
ii. Relation of NDVI with the moisture Index over the regions of different climatic types
iii. Sugarcane yield estimation and modelling based on NDVI data over Southeastern Brazil and
iv. Production and yield estimates of five major crops over Uganda

The importance of each of the above aspects in the context of assessing, monitoring and managing the natural hazards such as desertification, drought and risk management are discussed in detail in the individual sections

All case studies make use of time series of Normalized Difference Vegetation index (NDVI), an index of vegetation activity that can be derived from broad band measurements in the visible and infrared channels onboard satellite instruments and which is directly related to the photosynthetic capacity of plants [8]., Satellite sensors such as NOAA Advanced Very High Resolution Radiometer (AVHRR), Moderate Imaging Spectroradiometer (MODIS), Système Pour l'Observation de la Terre (SPOT) Vegetation etc. provide NDVI data on different intervals (8-day, 10-day, monthly and seasonal). NDVI value varies minus one (-1) to plus one (+1), whereby low NDVI values (< 0.2) reflect sparse vegetation, and higher NDVI values (> 0.4) reflect high vegetation densities. Several studies looked into the interannual variability and trends of NDVI in relation to meteorological parameters such as rainfall, temperature etc [9-11].The long time series of NOAA AVHRR has been widely used to relate the synoptic meteorology/climatology to understand the vegetation dynamics, vegetation response to climate and climate vegetation feedback mechanism [5, 12-14]. The studies of [9, 15-17] concluded that AVHRR NDVI is a valuable tool to monitor and asses large scale agricultural droughts. Other studies use NDVI time series derived from MODIS [18-20] or SPOT-Vegetation [21, 22].

In this work, the authors made use of NDVI imagery datasets derived from different satellites to relate the respective crop and climatic parameters to understand the sensitivity of NDVI with these parameters and further to use them for vegetation monitoring over large areas which contributes to risk management.

2. TIME SERIES ANALYSIS FOR DESERTIFICATION MONITORING

In the last decades, remote sensing technologies started to contribute enormously in documenting changes in land cover and monitoring desertification, drought and agricultural activities on regional and global spatial scales. Although desertification is highlighted as one of the most important global environmental issues, both desertification and greening processes have been reported on global scale [23-26]. These processes are related in many ways with other environmental issues, such as climate change and the carbon cycle, loss of biodiversity and sustainability of agriculture [27]. Also in South America environmental change is an important concern. In the last decades, South American ecosystems underwent important functional modifications due to climate alterations and direct human interventions on land use and land cover [28]. In South America, the main forest conversion process in the humid tropics in the period 1990-1997 was the clear-cutting of closed, open, or fragmented forest to make room for agriculture at a rate of approximately 1.7 million ha per year [29]. Apart from deforestation, also forest degradation occurs, a process leading to a temporary or permanent deterioration in the density or structure of vegetation cover or its species composition. Land degradation in arid, semi-arid and dry sub-humid areas is called desertification, and is the result of various factors, including climatic variations and human activities [30].

To determine desertification conditions, this case study focuses on vegetation dynamics in South America over a long time period based on a time series of low spatial resolution, high temporal resolution NDVI derived from SPOT-Vegetation, and to recognize to which extent this variability can be attributed to variability in rainfall, since rainfall is one of the most determinant factors of vegetation growth. In general, and especially in semi-arid regions, strong correlations between precipitation and the NDVI can be found. Therefore, the NDVI can be used as an indicator for vegetation status and vegetation response to precipitation variability. This study expands the analysis, as was performed on the Andes region [31] to South American continental level. In the first phase, trends of vegetation and precipitation indices are analyzed. In a next step, through correlation of NDVI and precipitation dynamics, these areas where the evolution of vegetation is not related to climate only and human induced impacts play an imperative role can be identified.

The time series of SPOT-Vegetation 10-daily composite NDVI data (April 1998 – March 2012) at 1 km resolution (http://www.vgt.vito.be/) was smoothed [32] and consequently synthesized to monthly images using the maximum value composite technique. Also 10-daily rainfall estimates at 0.25° resolution, available from the European Centre for Medium-Range Weather Forecasts (ECMWF) through MeteoConsult and the Monitoring Agricultural ResourceS (MARS) unit, were combined to retrieve monthly composites. The spatial resolution of the NDVI time series was degraded in order to fit the rainfall estimates using a weighted average approach. Many authors remove seasonality by integrating the data into annual values (e.g. [25, 28, 33]). In this study, in order to remove seasonal vegetation changes and thus facilitate the interpretation through the historical record, deviations from the 'average' situation were calculated for the NDVI time series using the Standardized Difference Vegetation Index (SDVI) [34] and for the precipitation time series using the Standardized Precipitation Index (SPI) [35]. In a next step, for each pixel a correlation analysis is performed on the monthly NDVI and SPI datasets, in order to identify the temporal scale at which the environment is most sensitive to precipitation anomalies (the so-called 'best lag').

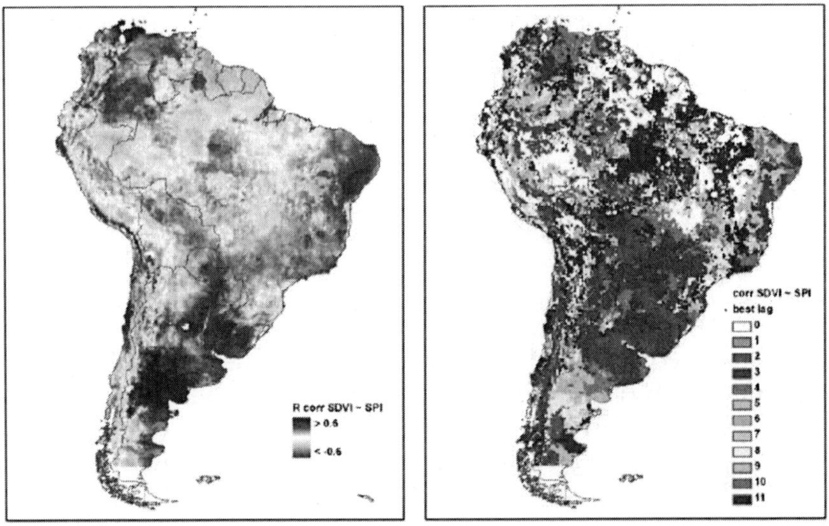

FIGURE 1. Correlation analysis between NDVI and SPI. Best lag expressed in months.

The results of the correlation analysis between NDVI and SPI are shown in Figure 1. Positively correlated (blue) areas suggest precipitation-vegetation coupling. In general, high positive correlation with best lags between 3 – 6 months are found in the semi-arid regions of South America, while weak (or even negative) correlation at all time scales was found in both the hot and humid zones and the deserts and high mountainous areas. These positive relationships between vegetation greenness and rainfall in drylands, where biomass production is determined by the amount of rainfall, and the opposite in humid and cold regions, where rainfall is not the limiting factor for vegetation growth, and deserts, where there is no rainfall at all, is consistent with findings of other authors, such as [33, 36].

In order to identify if a pixel is greening or degrading, and after identifying the best lag for each pixel, linear least squares trend analyses were performed on the SDVI time series and the SPI time series, taking into account the accumulated rainfall over the respective best lag. Only trends with Pearson correlation coefficients significantly different from zero (at significance level $p < 0.05$) are considered significant trends. Figure 2 shows the slope of the significant linear trends of vegetation greenness and precipitation anomalies, SDVI and SPI respectively. SDVI show slight but significant ($P<0.05$) positive trends in large areas in the northern part of South America, but significant negative trends in Argentina and the Peruvian coast. The results are comparable, but far more pronounced than results from annual series trend analysis, such as from [25, 28]. Also the SPI shows positive trends in the north-west of South America and the centre of Brazil. Negative precipitation trends are found in Argentina, the Peruvian coast and north-east of Brazil.

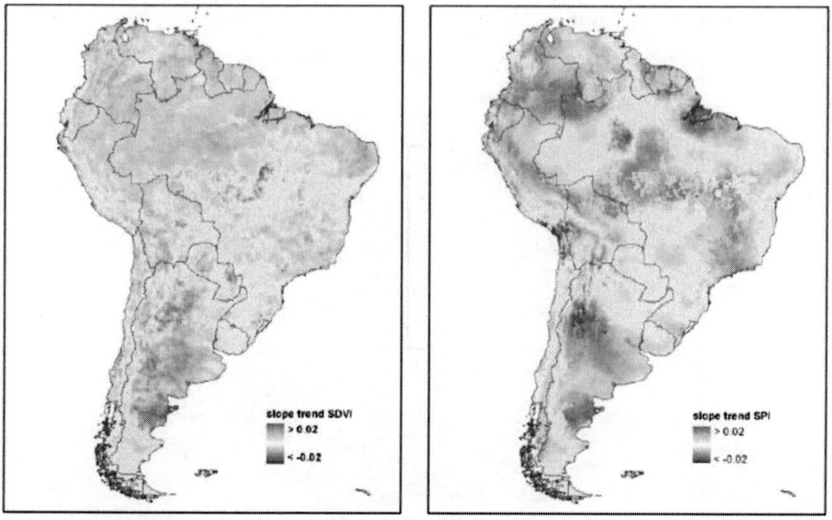

FIGURE 2. Slope of the trend analysis of SDVI (left) and SPI (right). Non significant trends are masked in grey.

Following the trend analyses, a decision tree approach is adopted in order to interpret the results. Five queries are stipulated:

1) Does the time series of SDVI show a significant trend?
2) Are significant trends in SDVI coupled to significant correlations between NDVI and SPI?
3) Are significant trends in SDVI linked to significant trends in SPI?
4) Does the SDVI show a positive trend?
5) Do trends in SDVI correspond to trends in SPI?

The first question is meant to identify the focus area, i.e. pixels where the de-seasoned vegetation index (i.e. SDVI) shows a significant linear trend over time. The second and third queries are used to identify areas that differ in their relationship between vegetation and precipitation trends, respectively, while queries 4 and 5 are meant to distinguish positive and negative trends that are linked to trends in precipitation. On the other hand, it is possible to identify regions where positive or negative trends in vegetation are not linked to changes in precipitation and other climate variables or human impact play an imperative role. In step 4, pixels are divided in greening or degrading pixels. In the last step, both the trends in SDVI and SPI are evaluated.

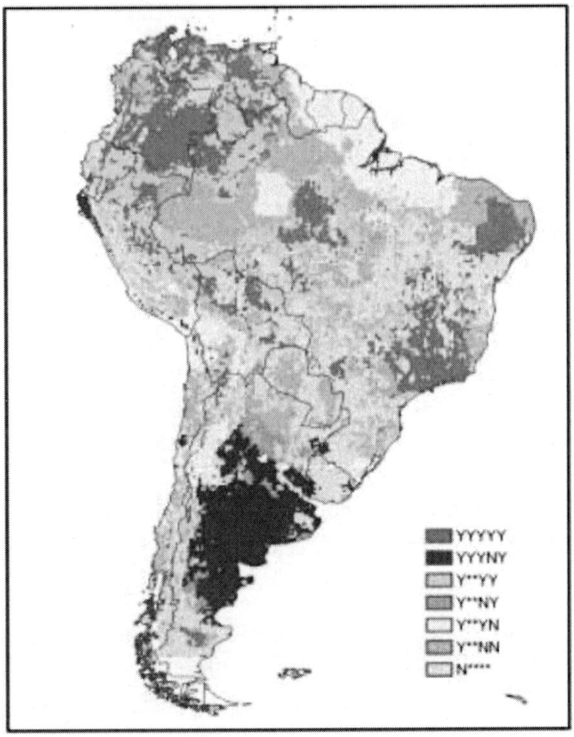

FIGURE 3. Result of the decision tree approach. Pixels are classified according to the 5 queries described in the text (where Y and N refer to a 'yes' or 'no' answer to the query, and * to either Y or N): green areas show an increase in SDVI, coupled to a significant positive trend in SPI (dark green) or not (light green); red areas show a decrease in SDVI, coupled to a significant negative trend in SPI (dark red) or not (light red); yellow areas show a positive trend in SDVI, but a negative trend in SPI; orange areas show a negative trend in SDVI, but a positive trend in SPI. Pixels without a significant trend in SDVI are masked in grey.

The results of the decision tree analysis are shown in Figure 3. Green classes show a significant positive trend in SDVI coupled to an increase in

SPI. Red classes show a coupling between a decrease in SDVI and SPI. The yellow and orange classes are classes where SDVI and SPI show an opposite trend. Argentina is clearly suffering from vegetation degradation linked to a decline in precipitation, which confirms the findings of [35]. The opposite is going on in Colombia and some parts of Brazil, where a process of greening seems to be linked to an increase in precipitation. Nevertheless, Brazil shows a patchy result, with some areas showing an increase in SDVI, although the precipitation decreases (yellow areas in Figure 3), and other areas showing a decline in vegetation, although precipitation increases (orange area in Figure 3), probably related to deforestation or forest degradation.

The resulting map can be used to estimate the coupling between vegetation (SDVI) and precipitation (SPI), shown in Figure 4. Three estimations are made: (A) a large estimation where significant trends in SDVI are coupled with trends in precipitation when SPI shows the same trend (significant or not); (B) an average estimation where significant trends in SDVI are linked to trends in precipitation when SPI shows a significant trend; and (C) a conservative estimation where significant trends in SDVI are linked to trends in precipitation when the SPI shows a significant trend and the correlation between NDVI and SPI is significant.

In estimation A, also weak trends in SPI are taken into account. These non-significant trends in precipitation are most probably not by itself responsible for significant trends in SDVI. Together with other variables like temperature change or human impact, these weak changes in precipitation might however give an extra impulse to greening or degradation processes. It is also possible that an increase in precipitation does not result in higher vegetation cover because the area is covered with climax vegetation or, the other way round, a decrease in precipitation does not result in further degradation because the area is already covered with minimal vegetation growth. Estimations A and B show little difference. The difference between estimation B and C is based on the significance of the correlation between NDVI and SPI, and shows larger differences, mainly in the greening pixels. In many pixels that show a positive trend in SDVI, vegetation greenness is not significantly correlated to SPI. It is therefore not certain that in these areas the increase in SDVI is coupled to an increase in precipitation.

From the conservative estimation, we can conclude that in 8% of South America, vegetation degradation is coupled to a significant decrease in the amount of precipitation in the last 14 years. Our results corroborate with the findings of [37]. In contrast, in 18% of the subcontinent, vegetation greenness has significantly increased over the last 14 years, coupled to an increase in precipitation. For 46% of the study area, significant degradation or greening processes could not be linked to changes in precipitation over time, indicating human impact or the influence of other climatic factors, such as temperature. Finally, and without taking into account the link with trends in precipitation, 40% of the subcontinent is showing an increase in photosynthetic activity over time, while desertification is taking place in 32% of the area.

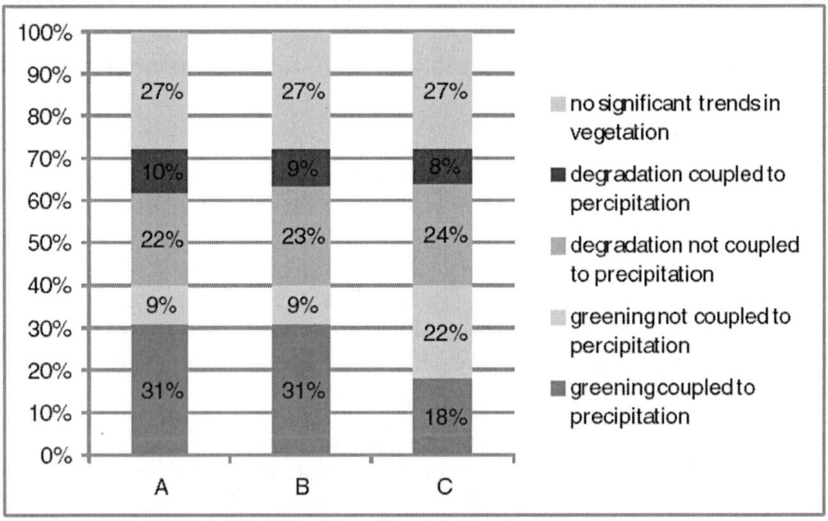

FIGURE 4. Large (A), average (B) and conservative (C) estimation of the linkage of greening and degradation to changes in precipitation over time. In case of estimation B and C, the yellow areas in Figure 3 are included in the light green fraction, while the orange areas in Figure 3 are included in the light red fraction.

3. TIME SERIES ANALYSIS FOR DROUGHT MONITORING

Drought is a recurrent feature of the Indian climate and usually begins at any season and can prolong for many years. The study of drought characteristics is to ascertain the spatial and temporal distribution of droughts, synoptic meteorological conditions associated. The definition of drought mainly depends on the precipitation deficiency. Studies on droughts in India have been reported by many scientists based on the rainfall anomalies over a particular region [38-40]. According to India Meteorological Department (IMD) guidelines, drought is defined as the consequent rainfall deficiency (below 19% of normal) for a period of 2 consecutive weeks. However, this criterion varies from country to country, based on the meteorological/climatological conditions such as percentage of moisture present, land topography etc.

Drought is of different types including meteorological droughts, agricultural droughts, hydrological drought and socioeconomic droughts. These are based on the variations of rainfall, crop water, surface water and economic conditions respectively. These droughts show variation with respect to the climatology which prevails of that region. The climatology of a region is a replica of the severity of drought. The climatology can be derived from the Thoronthwaite Climate System by taking the inputs of rainfall and water need. The model gives the amount of moisture (annual/seasonal) from which the classification of climate can be studied.

In view of the above, the present study focuses in obtaining the climates in different parts of Karnataka state which is located at the western half of the Deccan Palateau of India. The Moisture Index values which are the basis for delineating climatic type were compared with the AVHRR NDVI to understand the drought climatology in different test regions of Karnataka, India (Figure 5).

The rainfall (P) and potential evapotranspiration (PE) data for the period 1982 to 2000 was downloaded from [41]. This data is based on the global rainfall and temperature (PE can be calculated from temperature) data sets of Climate Research Unit, University of East Anglia, United Kingdom. This was averaged for all the districts of Karnataka state till the year 2000 and uploaded website.

Taking the inputs of P and PE, we run awater balance model and derived the monthly Aridity Index (I_A) and Humidity Index (I_H). Moisture Index (I_M) which is the basis to tell the climatology of a region can be obtained by subtracting I_A from I_H.. Table 1 below shows the infered climate types based on $I_M$as per Thoronthwaite Climate Approach [42, 43].

The selected districts for the study are Chikkamagaluru, Belgaum, Chamrajnagar and Gulbarga of which climates are Humid, Dry subhumid, Semiarid and Arid. A comparative study was made with the seasonal values of IM and NDVI using time series and correlation analysis.

FIGURE 5. Study area of the present investigation

TABLE 1. Classification of climatic types based on Thoronthwaite Approach

Moisture Index(IM)	Climatic Type	Notation
100 & above	Perhumid	A
80 to 100	Humid	B4
60 to 80	Humid	B3
40 to 60	Humid	B2
20 to 40	Humid	B1
0 to 20	Moist Subhumid	C2
-20 to 0	Dry Subhumid	C1
-40 to -20	Semiarid	D
-60 to -40	Arid	E

Table 2 shows the climatology of the four selected test sites during the period of 1982 to 2000. The overall climate of the test regions for the study period represented the humid, dry subhumid, semi arid and arid for Chikkamagaluru, Belgaum, Chamrajanagar and Gulbarga districts respectively.

TABLE 2. Climatic types of test regions from 1982 to 2000

Year	Climate Type			
	Chikkamagaluru	Belgaum	Chamarajanagar	Gulbarga
1982	C1	C1	D	E
1983	B1	C2	D	E
1984	C2	D	D	E
1985	C1	D	D	E
1986	C2	D	D	E
1987	C1	D	D	E
1988	B2	C1	D	E
1989	C2	D	D	E
1990	C1	D	D	E
1991	B1	C1	D	E
1992	B2	C2	D	E
1993	B1	B1	C1	E
1994	B2	C2	D	E
1995	C2	C1	D	E
1996	B1	D	D	E
1997	B1	C1	C1	E
1998	B2	C2	C2	E
1999	B1	B1	C1	E
2000	B1	C2	D	E

3.1. Chikkamagaluru (Humid Region)

The climatic types of Chikkamagaluru were dominated by humid type which is followed by the dry sub humid type. Chikmagalore (Humid region) recorded a maximum I_M of 450 during the year 1994 (Figure 6). The variation of I_M shows that it has increased from June to August in all years and recorded a comparative less value in the month of September. In a similar way, the NDVI progressed from June to September in all years with a maximum value of 0.55 for September, 1992. It is conspicuous that the variability in NDVI is more than I_M and the trends of I_M and NDVI were increasing. The correlation coefficient of these two indices is +0.08 which is very poor. The studies of [38, 39] also suggested that understanding the relation of NDVI with rainfall and its by-products is a very difficult task in humid regions. Since the plenty of moisture is already available in the soil, the vegetation can utilize the moisture for its growth and in such case it may not directly/immediately dependant on rainfall derived indices in humid regions.

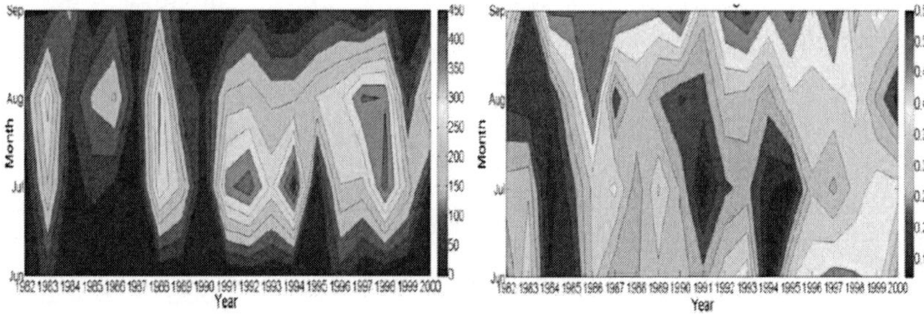

FIGURE 6. Variation of IM and NDVI – Chikkamagaluru

3.2. Belgaum (Dry Subhumid)

The climate of Belgaum varied from semi arid to 1st humid type during the study period. 10 years of the study period show the subhumid climates followed by semi arid in 7 years. The variation of $I_M$with NDVI over Belgaum unraveled the low variability of I_M associated with high variations of NDVI (Figure 7). The maximum NDVI is 0.55 that is during September of 1984 and 1992 linked with the very low values of I_M. The I_M values were around zero during all the years of study period in June and July months countered by the low values of NDVI. There is no sea – saw relation found between I_M and NDVI over this region. The correlation in this case was also found to be very poor and insignificant.

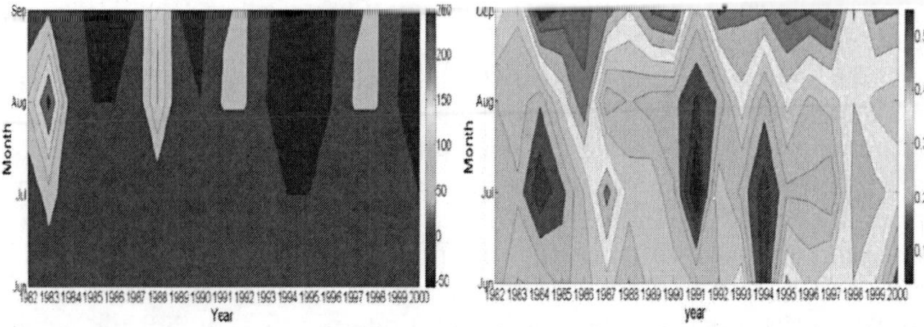

FIGURE 7. Variation of I_M and NDVI – Belgaum

3.3. Chamrajnagar (Semiarid)

From Table 2, it is noticed that Chamrajanagar shows the arid climate category in 15 years of the study period. The years such as 1993, 1997, 1998 and 1999 displayed subhumid climates. The climatic types of this region are the good representative of climatic features over the study period. In the case of semi arid region of Chamrajnagar, the variability of I_M and NDVI is better as compared with the humid and dry subhumid regions (Figure 8). The overall I_M is varied from -10% in the year 1984 to a maximum of 44% during the year 1998 for the south west monsoon season. Accordingly NDVI, also varied from 0.26 to a high value of 0.42 in the year 1996. The values of NDVI are less in this region than previously mentioned areas but the trends of I_M are NDVI were positive with slopes of 0.824 and 0.002 for I_M and NDVI respectively along with the standard deviations of 11.5 and 0.05. The correlation of these two data sets is 0.42 at 0.05 level of significance which infers the good agreement of I_M and NDVI. From this analysis, it can be noticed that the deficiency of moisture which is represented by I_M in this study was well reflected by low NDVI values and adequate moisture conditions are supported by the moderate vegetation conditions.

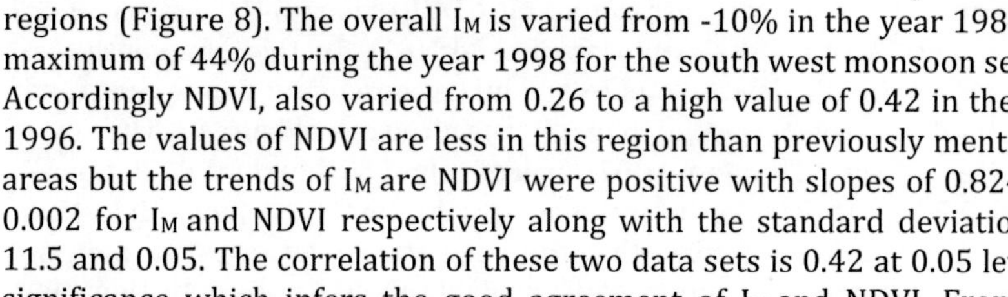

FIGURE 8. Variation of I_M and NDVI – Chamrajanagar

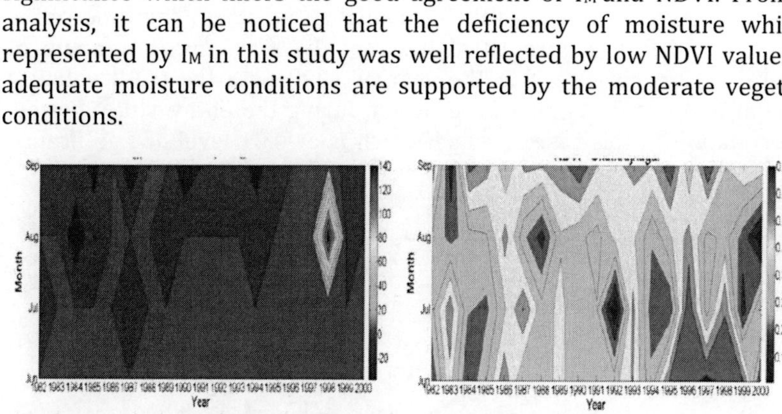

FIGURE 9. Variation of I_M and NDVI – Gulbarga

3.4. Gulbarga (Arid)

All the years of study period are dominated by the arid category for this region. The comparison of $I_M$and NDVI yielded good results in the arid region of selected test sites (Figure 9). The interannual variability was found to be very high both in I_M and NDVI where as I_M varied from -77% during August of 1982 to zero value during September month in the years 1983 and 1992 with the corresponding NDVI values of 0.055 and 0.127 respectively. The time series plot for I_M and NDVI for the total south west monsoon season display the one to one linear agreement where the trends of both were highly increasing than other test regions (Figure 8). The slopes of the trends were 1.68 and 0.007 respectively. The standard deviations of 14 and 0.05 infer that the interannual variability of I_M is more than NDVI from which it can be noticed that the vegetation over a region may not respond immediately to the rainfall/available moisture despite there is a dependence of vegetation on rainfall/available moisture. The time series plot of I_M and NDVI (Figure 10) shows that the maximum of amount of moisture of-12% have seen in the year of 1998 with the NDVI value of maximum NDVI of 0.269 that is recorded during the entire study period. The correlation in this case is +0.64 which is at 0.01 level of significance which shows the strong agreement between I_M and NDVI.

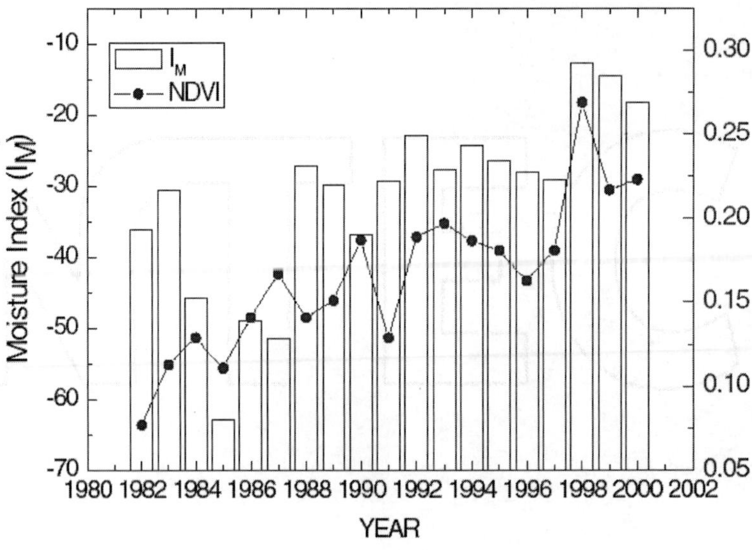

FIGURE 10. Time series of I_M and NDVI for southwest monsoon

The study commenced with the retrieval of I_M values from the water balance model on monthly basis over different selected test regions those represent the various climatic types such as humid, dry sub humid, semi arid and arid. The knowledge of different climatic types enables us to understand the climatology of the test regions during the study period. Since these climates were derived from the $I_M$values, they replicate the status of the moisture content available which is very essential input to decide the crops fate. The comparison study of I_M with the satellite derived NDVI shown very interesting features of sensitivity of NDVI with I_M over different climatic types. The study inferred the poor correlation such that no linear and significant relation of I_M with NDVI over humid and dry subhumid regions. The reason for this could be the plenty of available moisture over these regions and even temporary perturbations of land surface conditions may not affect the crops/agriculture. The relation grown up to strong when the comparison closes from semi arid to arid regions. Especially, Gulbarga, arid region displayed very strong relation of I_M with NDVI which unraveled the poor/good vegetative conditions associated with low/high values of I_M. The correlation of +0.65 is a good supporting factor to say that the relation is substantial. The overall analysis of the present study suggested that the relation of I_M with NDVI is very strong and it is of immense use for the studies of drought monitoring in the arid areas as compared with the other climatic types.

4. QUANTIFICATION OF SUGARCANE CROP PRODUCTIVITY: A STUDY CASE IN SOUTHEASTERN BRAZIL

Agriculture represents an important segment of the economy of Brazil. Over the past 30 years, Brazilian agricultural growth and development has been guided by policies and technologies based on research for development. Remote sensed imagery plays an important role in agricultural crop production over large area, quantitatively and non-destructively, because agricultural crops are often difficult to access, and the cost of ground estimating productivity can be high. The recent development of GEONETCast–EUMETCast data has allowed us to obtain frequent and accurate measurements of a number of basic agrometeorological parameters (e.g. evapotranspiration, surface albedo, surface temperature, solar radiation, rainfall etc.). The GEONETCast–EUMETCast real-time and on-line data dissemination systems represent global network of satellite-based data dissemination systems designed to distribute space-based, air-borne and in situ data, metadata and products to diverse communities.

To determine agriculture productivity, this case study aimed to develop a GEONETCast-EUMETCast product-based method of estimating the productivity of sugarcane using an agrometeorological spectral model. The study was carried out in the Municipalities of Barretos and Morro Agudo, located in the state of São Paulo, Southeastern Brazil (Figure 11). The analysis was performed for 2009/2010 and 2010/2011 year's crop.

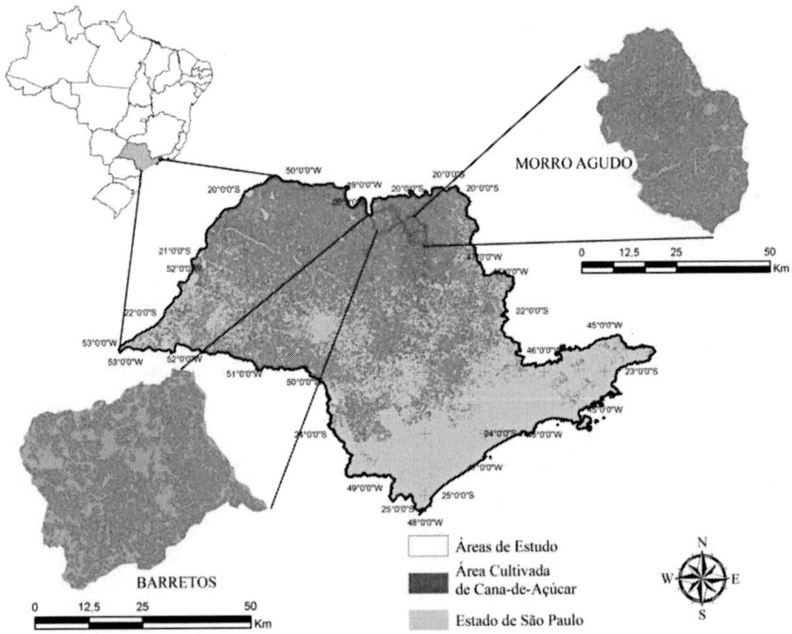

FIGURE 11. Spatial variability of crop yields (2010/2011) of Barretos and Morro Agudo in Sao Paulo, Brazil.

The values of sugarcane parameters used such as Respiration Factor (RF) (0.5 for temp. ≥ 20°C and 0.6 for temp <20°C); Agricultural Productivity Factor (APF) (2.9), Yield Response Factor (Ky) and Crop Co-efficient (Kc) were taken from [44-46]. The EUMETCast service is installed at Laboratory of Analysis and Processing of Satellite Images (LAPIS) at Federal University of Alagoas (UFAL). The remote sensing data of NDVI S10, Production of Dry Matter (DMP) are available at the website http://www.lapismet.com and SPOT Vegetation indices of VITO were collected form http://free.vgt.vito.be/.

This application proposes to test a remote sensing approach to quantify estimates of sugarcane productivity over the Coruripe municipality with the Integrated Land and Water Information System, (ILWIS, 3.7.1) GIS software. ILWIS was used to compute sugarcane crop estimates for each pixel in NDVI DMP and ETp images by applying radiative, aerodynamic and energy balance physics in 7 computational steps. These images are currently provided over both daily and 10 day composites at about a 3km and 1km spatial resolution, by EUMETSAT and VITO respectively.

Step 1: Input NDVI and DMP databases using algorithm adapted from GEONETCast Toolbox

To implement a methodology for the ingestion of both the NDVI and DMP databases (raster) into ILWIS, specific routines of GEONETCast–toobox are adapted to import the datsets. For the ingestion procedure, based on a GIS approach using open source components, it requires additional work on corrections using overlays (Status map and LOG–file) to mask all appropriate areas of sugarcane crops over the Coruripe municipality.

Step 2: Computation of Fractional Vegetation Cover (FVC) from NDVI

For each pixel, the NDVI is converted to Fractional Vegetation Cover (FVC) by means of the the formula of [47]. The FVC is the one biophysical parameter that determines the contribution partitioning between bare soil and vegetation for surface evapotranspiration, photosynthesis, albedo, and other fluxes crucial to land–atmosphere interactions.

$$FVC = 1.1101 * NDVI - 0.0857.$$

Step 3: Computation of Leaf Area Index (LAI) from FVC

For each pixel, the FVC is converted to Leaf Area Index (LAI) by means of the formula of [48]. The LAI, defined, as the total one-sided leaf area per unit ground area, is one of the most important parameters characterizing a canopy. Because LAI most directly quantifies the plant canopy structure, it is highly related to a variety of canopy processes, such as evapotranspiration, interception, photosynthesis and respiration.

$$LAI = -2 \mathrm{Ln}\,(1 - FVC).$$

Step 4: Computation of growth factor from LAI

[49] developed a simple approach for deriving growth rate equation from LAI. Experimental evidence indicated that the growth rate of several agricultural crop species increases linearly with increasing amounts of LAI, when soil water nutrients are not limiting [46]. The following equation is used:

$$CGF = 0.515 - e^{[-0.667 - (0.515*LAI)]}$$

where *CGF* = Corrected Growth Factor. Experimental evidence indicated that the growth rate of several agricultural crop species increases linearly with increasing amounts of LAI, when soil water nutrients are not limiting.

Step 5: Computation of maximum yield potential (Yp)

The final equation that was used to derive maximum yield potential (Yp) includes evaporative fraction corrected growth factor (CGF), respiration factor (BF), agricultural productivity factor (APF) and production of dry matter (DMP) product.

$$Y_p = CGF*BF*APF*DMP$$

where Yp is the maximum yield potential (kg ha^{-1}).

Step 6: Retrieval of evapotranspiration (ET_p) via Land Surface Analysis –Satellite Application Facility (LSA SAF) ET_p product

The crop coefficient is defined as the ratio of crop evapotranspiration, ET_r, to reference evapotranspiration, ET_p. K_c is crop specific and ranges from zero to over unity, depending on the crop growth stage. Crop evapotranspiration at any time during the growing season is the product of reference evapotranspiration and the crop coefficient.

$$ET_r = ET_p * K_c$$

Crop coefficients was developed for nearly all crops by measuring crop water use with lysimeters and dividing the crop water use by reference evapotranspiration for each day during the growing season of 2009/2010 [50].

The recent development of LSA–SAF products has allowed us to obtain frequent and accurate measurements of a number of basic agrometeorological parameters (e.g. surface albedo, surface temperature, evapotranspiration). The satellite estimated agrometeorological parameters have several advantages compared to conventional measurements of agrometeorological data in ground meteorological network.

Step 7: Estimation of sugarcane productivity

The sugarcane yield estimation model over the growing season, on a biweekly basis, is accomplished by using an agrometeorological model integrated to ILWIS according to [46]:

$$Y_e = Y_p[1 - ky(1 - ET_r/ET_p)]$$

where *Ye* is the estimated yield (kg ha^{-1}), *Yp* the maximum yield (kg ha^{-1}), *ky* the yield response factor; ET*r* the actual evapotranspiration (mm) and ET*p* the maximum evapotranspiration (mm). Maximum yield (*Yp*) is established by the genetic characteristics of the crop and by the degree of crop adaptation to the environment.

The resulting map of the estimated yield (*Ye*) is clipped to mask the Coruripe municipality boundaries in the State of Alagoas, Brazil. To establish correct coordinates, Map calculation within the ILWIS is used to implement this procedure. Flow diagram of methodology of quantifying sugarcane productivity via satellite products is shown in Figure 12.

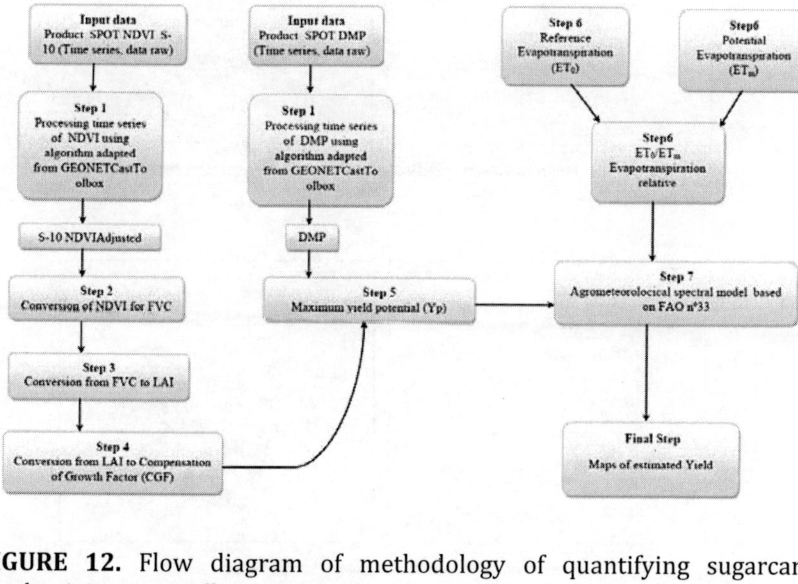

FIGURE 12. Flow diagram of methodology of quantifying sugarcane productivity via satellite products.

Figure 12 shows the spatial variations in sugarcane production over the Barretos and Morro Agudo municipalities for 2009/2010 and 2010/2011. The figure clearly indicates high spatial patterns in yield variability. This could be due to the mixing of significant fraction of observed pixels for the "arable pixel" and "non-arable pixel" within the municipalities. The quantified results give sugarcane yield mean range of 50 to 135 Ton ha^{-1}.The results obtained here represents a first step towards an operational use of ILWIS tools in Brazil using NDVI S-10, DMP SPOT and ETo for operational estimating of sugarcane productivity. Overall, the model was able to identify (Figure 12) and quantify (Table 3) the spatial variability of agricultural production over the municipalities analysed. Therefore, the methodology is useful for developing estimates of operational support for the sugarcane productivity [51].

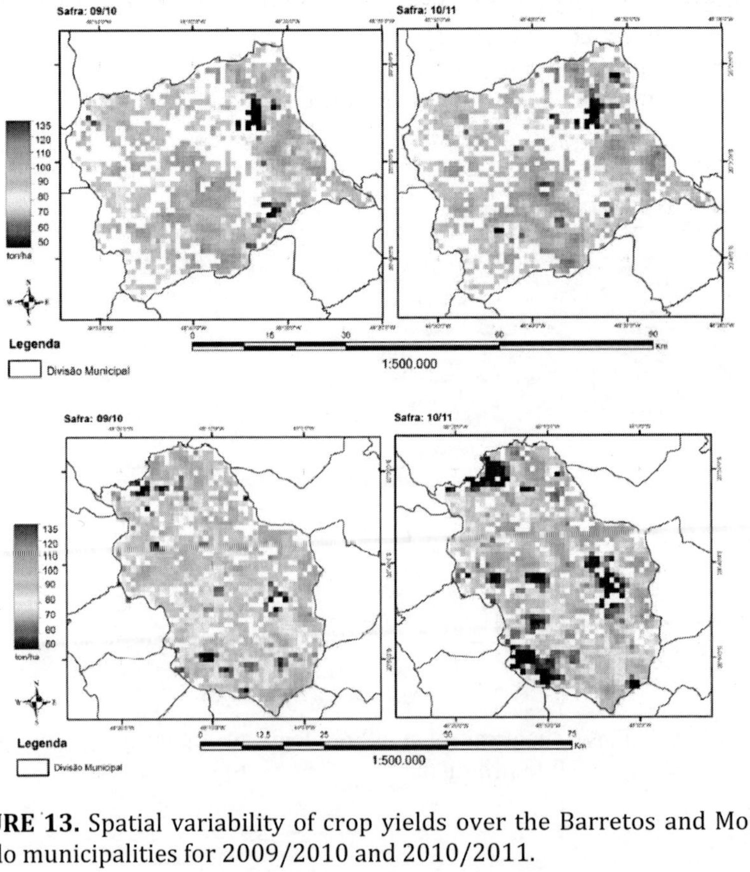

FIGURE 13. Spatial variability of crop yields over the Barretos and Morro Agudo municipalities for 2009/2010 and 2010/2011.

TABLE 3. Comparison between the productivity of sugarcane using an agrometeorological spectral model and harvested crop yield from National Food Supply Company (CONAB).

Mean Productivity (t/ha)		
Barretos	Crop	
	2009/2010	2010/2011
Agrometeorological model	93,15	93,31
CONAB	96,21	93,29
Mean Productivity (t/ha)		
Morro Agudo	Crop	
	2009/2010	2010/2011
Agrometeorological model	87,96	84,10
CONAB	93,60	87,25

5. TIME SERIES ANALYSIS FOR AGRICULTURE MONITORING: UGANDA

The economy of Uganda and its development goals are heavily premised on agriculture. Over 79% of the households are engaged in agriculture while 73% are directly or indirectly employed in the agricultural sector. Uganda's agriculture is however almost entirely rain-fed and very susceptible to climate risks. Studies indicate that Uganda's agricultural sector will be adversely affected by climate variability and projected climatic changes making real time monitoring of crop growth and crop productivity very important for better adaptation to climate variability and climate change. Quantitative analyzes reveal that the agricultural sector in Uganda needs to grow at an annual rate of 7% to effectively contribute to national development. Currently, the rate of agricultural growth in Uganda is below the population growth. With the threats of climate change and variability, Uganda needs to among other things harness geo-information technologies (remote sensing and geographical information systems) to improve agricultural productivity.

Remotely sensed images are powerful tools in monitoring crop productivity and yields. In most developed countries and emerging developing countries like Brazil and India, remote sensing has been greatly harnessed to plan for agriculture production, monitor crop growth and estimate yields. This is paramount in the sense that timely interventions can be taken and obviates possibilities of famine and food insecurity. Although there have been strides taken to improve the utility of remote sensing in the agriculture some developing countries, a lot remains to be done to make it more efficient, relevant and more productive. An investigation of the causative factors of the low utility and uptake of remote sensing in the agricultural sector in Uganda implicates a number of factors ranging from low capacities to expensive images. Recent developments have however extended numerous opportunities in utilizing remote sensing in the agricultural sector.

The onset of utilization of remotely sensed techniques in Uganda was in the early 1990s spearheaded by the National Biomass Project and focused largely on land use and land cover mapping. The activities of the National Biomass Project were later taken on by the National Forestry Authority (NFA) but the domains and scope remained largely the same with more focus on land use, land cover and related aspects being given priority. Apart from the NFA, academic institutions of higher learning and to some extent some research institutions like the National Agriculture Research Laboratory Kawanda (NARL) and the National Environment Management Authority (NEMA) have some remote sensing application either for teaching or research. In general, the remote sensing applications in Uganda in the agricultural field are scattered and more project based. This is partly due to the fact that there is lack of a government agency with a clear mandate to spearhead and propel the utility of remote sensing application in the country. Nevertheless, some efforts through the government cooperation with UN agencies such as the FAO regularly provide some information analyzed at the regional level for early warning in the agricultural sector. Some of the historical constraints to efficiently harnessing remote sensing in natural resource management in Uganda are generally those also experienced in other developing countries in Africa including the high costs of imagery data, processing software, coarse resolution of images, inadequate physical and human capacities and weak institutions. To-date, most of the issues to do with data costs and software have been significantly resolved with many freely available images, open source versatile software or special low for developing countries on commercial software. The contemporary challenge now is more of institutional/agency capacities, human capacities and policy environment for enhancing the utilization of remote sensing in the country.

A range of great opportunities, hitherto unavailable exist now for effectively using remote sensing in agriculture and natural resource management, notably through; (a) datasets disseminated through the Geonetcast Platform (b) freely available and downloadable datasets (c) open source softwares and low cost commercial softwares. Details of the Geonetcast is fully described in various sources (e.g. [48-50]). In brief, its a low cost facility which enables dissemination of near real time satellite imagery data. It is part of the emerging Global Earth Observation system of Systems (GEOSS), led by the Group on Earth Observation (GEO), for environmental analysis [54]. The Geonetcast does not require internet connectivity which is always a major constrain in developing countries and the data is disseminated at a very high temporal resolution through a ground recieving station, making monitoring easy. The facility streams diverse datasets which can broadly be used in environmental monitoring covering agriculture, water, soils, fire forestry etc. The data can be processed using the ILWIS software, where a specific toolbox has been developed.

In this case study, we demonstrate the utility of relatively low spatial but high temporal resolution satellite images from earth observation systems in monitoring and assessment of agricultural productivity in Uganda. There are two main inputs i.e. production data and remotely sensed data. Production data was obtained online from the FAOSTAT [55]. We extracted the annual yield, production and harvested area of the top five crops produced in Uganda according to FAOSTAT; (1) plantain/banana (2) cassava (3) sweet potatoes (4) sugarcane and (5) maize for 10 years spanning from 2001 to 2010. Bananas/plantain (*Musa spp*) are largely grown in central, western and eastern (highland areas) parts of Uganda. As perennial crops, banana are year round crops. Cassava (*Manihot esculenta*) is an important food security crop in Uganda with the largest production coming from eastern and northern Uganda. Cassava accounts for approximately 13% of the daily caloric intake in Uganda. Cassava is commonly planted in the first season which is around February-march in most parts of the country and its also a perennial crop. Sweet potatoes (*Ipomoea batatas*) are also a food security crop in Uganda grown largely in the mid to high altitude regions of the country (1000-3000 meters above sea level). They are annual crops grown twice a year with the first season between February and June, while the second season stretches from September to November. They thrive well in the deep volcanic soils of Southwesten Uganda and Eastern Uganda. Sugarcane (*Saccharum officinarum*) in Uganda is largely grown on large plantations mainly in the near east and western Uganda. There are also a couple of out growers who are supported by sugar companies. It is mainly an income generating perennial crop. Maize (*Zea mays L.*) is grown in almost every part of the country and is a major staple food crop. It is an annual crop grown twice a year (March to June and September to November) in areas of the country where biophysical conditions are supportive. To ease the analysis, the production and yields for the five crops were compounded into one annual value.

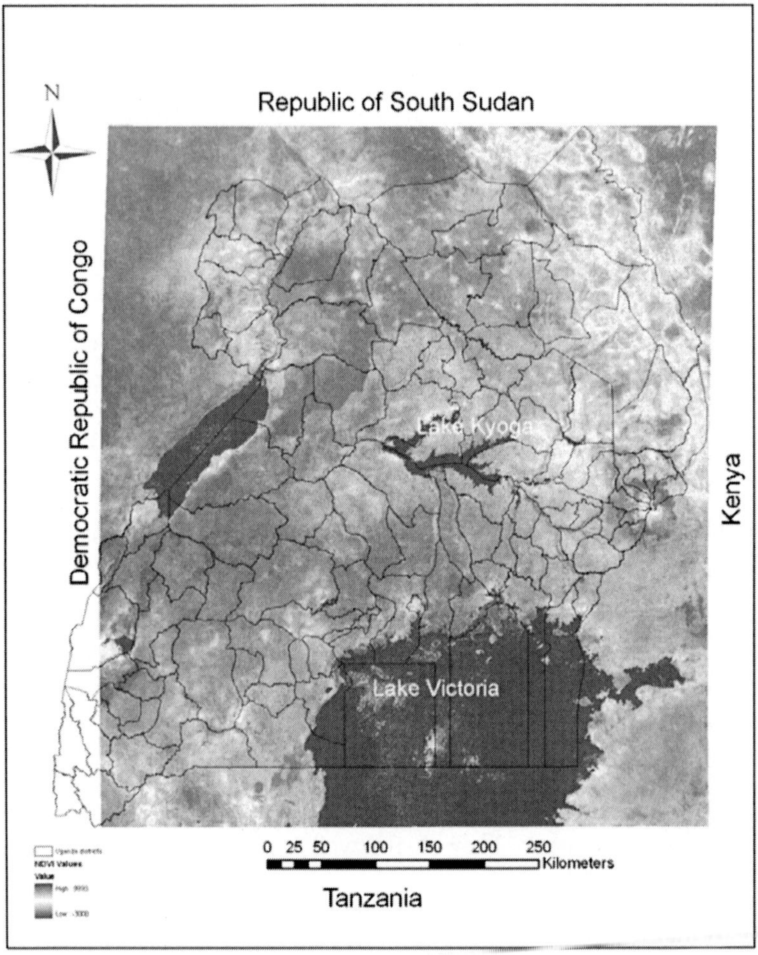

FIGURE 14. Scope of study

The results on harvested area, production and yields are shown in Figures 15, 16 and 17 respectively.

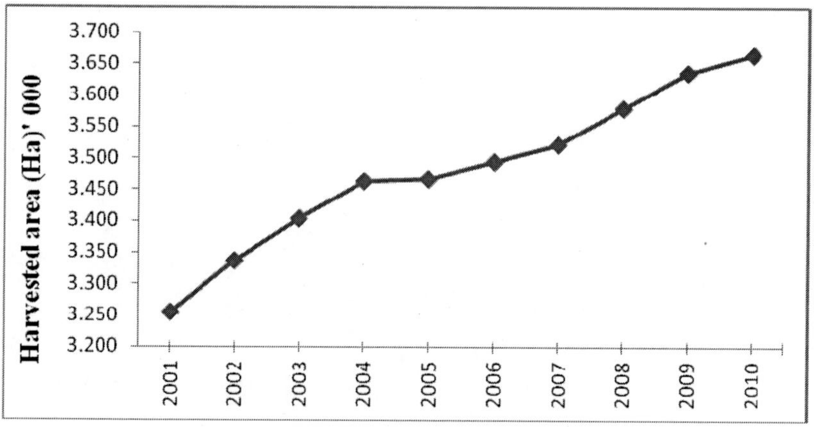

FIGURE 15. Harvested area of five crops between 2001 and 2010.

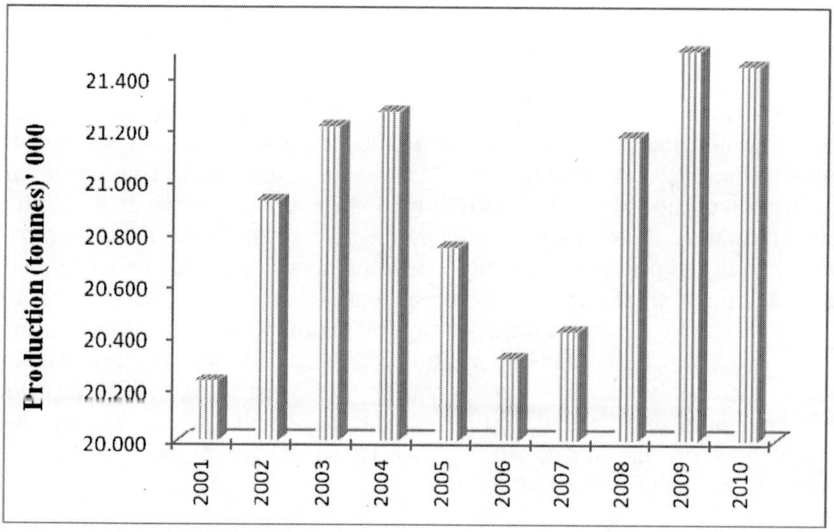

FIGURE 16. Production (compunded) trends of five selected crops between 2001 and 2010 compounded

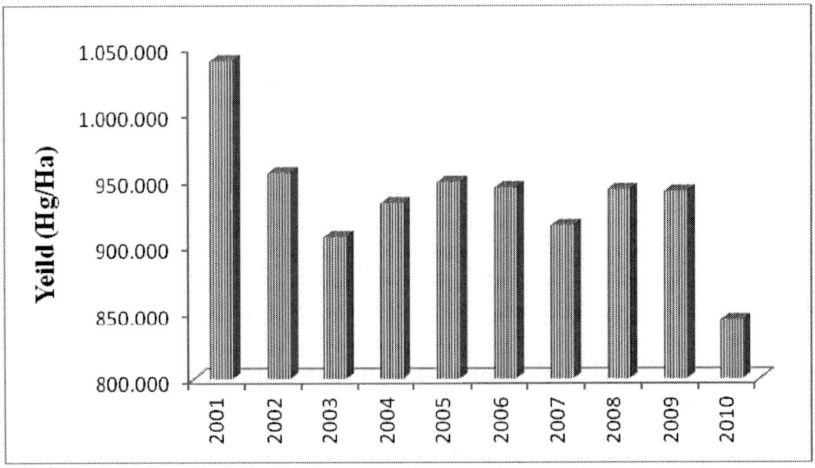

FIGURE 17. Yields (compounded) of five major crops between 2001 and 2010

The results based on the three factors i.e. area harvested, production and yields do not depict a definitive trend. In terms of harvested area, there is an increasing trend, implying that more areas are being converted for cultivation of the specified crops (Figure 15). Production between 2001 and 2010 has modestly increased. Yields per hectare are however more variable and actually show and generally declining trend. Bearing in mind that production is increasing, it becomes explicit that the increments in production are related to extensification rather than intensification. In most cases, extensification entails conversion of ecologically sensitive and fragile areas such as wetlands or reclamation of forest area which has its environmental implications. Subjected to a statistical analysis, the results revealed a strong and positive correlation between the yields and production area ($r^2=0.52$, $p<0.05$).

Remote sensing analysis was on the MODIS NDVI data, which has a spatial and temporal resolution of 250 m and 16 days respectively. For each year 23 images are available in a decal arrangement. For the 10 year period, we downloaded a total of 230 images in HDI format and processed them in ERDAS Imagine where file format conversions were undertaken and later ILWIS for arithmetic analysis. Individual images (23 decades) for each year were stacked to generate a single profile for each year. Relevant statistics such as the mean, standard deviation, coefficient of variation were late extracted. The spatial distributions of average NDVI for selected years are shown in Figure 18, whileFigure 19 gives the temporal average NDVI dynamics for the 10 years. Mean average NDVI value is 0.56. In spatial terms, the southern part of the country registers higher NDVI values than the northern part. This is not surprising in light of the coverage in terms of natural cover and the crops grown which significantly entail banana and a range of annual crops grown in two seasons. The North Eastern part is particularly poorest in terms of annual average NDVI values. Understandably it is a semi arid region and generally more tailored to livestock enterprises than cropping enterprises. Annual NDVI values for the whole country were subjected to a correlation analysis with production and yield data, resulting into poor and insignificant correlations (r^2=0.19 to 0.2.1). The low coefficient are partly explained by the fact that some crops like sugarcane are eother irrigated or are grown in areas almost permanently under water (wetlands). However when the data was collapsed into the growing seasons and the water bodies excluded from the analysis, better and significant correlations were obtained (r^2 0.46 to 0.61, $P<0.05$) demonstrating the efficacy of using NDVI for crop monitoring and yield prediction. In light of the expected variability and changes in climate, coupled with the availability of data in real time, the NDVI analysis represents a great potential in sustainable adaptation where from both a policy perspective and direct intervention. This has positive implications in timely provisioning of information to farmers, adaptation to climate change and variability as well as enabling science based policy options for appropriate interventions. Specific prediction coefficients for different crops and regionalized to the climatic conditions can be helpful to local governments where timely interventions can obviate social instability related to crop failures. On the other hand, predictions of higher yields can also enable relevant agencies to solicit for markets for the produce, improving the welfare and livelihood of the farmers, who in the context of Uganda are largely small holder farmers. All these can only be realized if there is a good policy framework that ties all the relevant pieces in the chain i.e. science, production, markets and institutions.

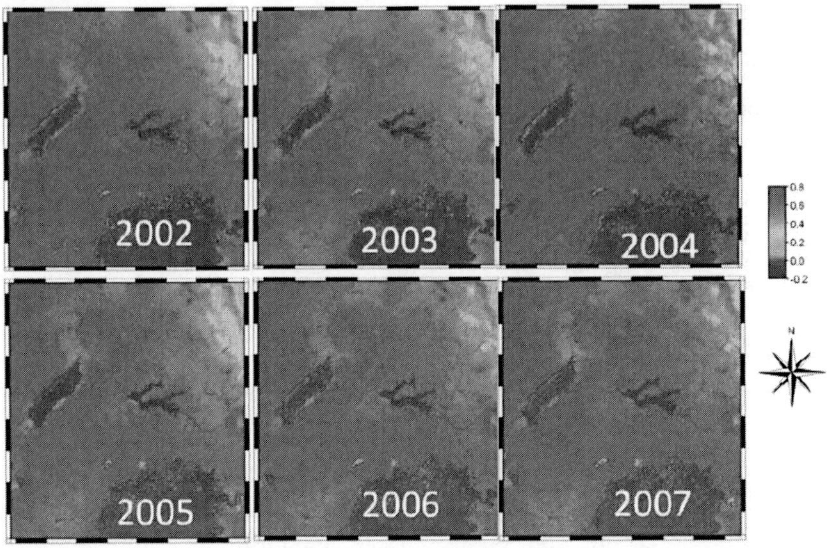

FIGURE 18. Average annual NDVI for selected years

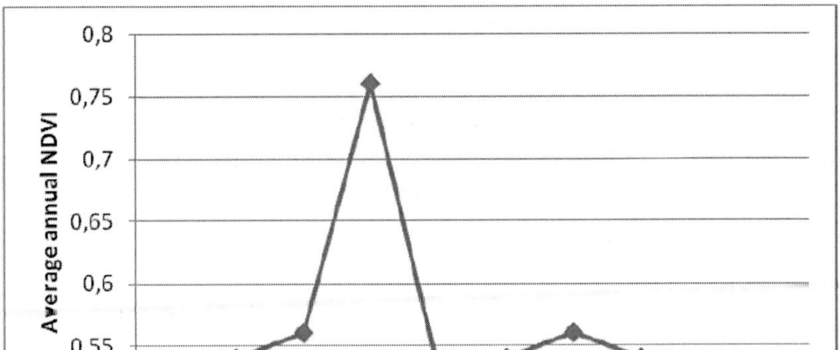

FIGURE 19. Variation of annual NDVI values between 2001 and 2010

6. OUTLOOK AND CONCLUSIONS

We have provided four multi-disciplinary case studies on the power of using remote sensing technologies, and more specifically time series analysis of low resolution satellite derived indicators, for monitoring and analysing land cover changes, desertification, drought and agricultural activities on different spatio-temporal scales. Generating this information at finer temporal resolutions is crucial for reducing risks to disaster, preparedness and formulation of strategies for better adaptation to climate change particularly the increasing dramatic hydro-meteorological events in developing and emerging countries.

This chapter provides a variety of methodologies of processing chains over satellite data, allowing the monitoring of areas subject to or in risk of desertification and land degradation processes. This chapter provides new insights related on the use of remote sensing data for climate (change) impact monitoring, which will contribute to the advance of warning systems and adaptation measures in developing and emerging countries. The focus of future activities should however focus on institutional support and capacity building for impact assessments for Africa, South America and India. The importance of training and joint cooperation with local providers and users cannot be overestimated.

One of the most robust, multi-purpose and yet simple remote sensing index is the NDVI. NDVI imagery data is widely available for immediate use at almost no cost. This has been given emphasis in this chapter through demonstration of its utility in various environmental and production domains. The section 2 of the chapter mainly emphasizes the relation between trends of vegetation greenness and rainfall over a long term period, taking into account the time lag between rainfall and vegetation response. As a result, areas of greening or degradation can be identified, and the process can be linked or not to changes in precipitation.

The section 3 of the chapter tells the mode of relation between NDVI and moisture index over different climatic regions. The relation was found to be poor over humid and dry subhumid regions where as it is improving in semi arid and arid regions. The relation of above cannot be taken as granted in the humid regions though it is implicitly understood that NDVI maintains positive relation with IM. The study infers that the NDVI and IM relations cannot be used to characterize the drought over humid regions but can be taken as an indicator in arid and semi arid regions. This is particularly relevant for adaptation purposes in semi arid regions which cover big chunks in Africa, India and some parts of Southern America.

Section 4 of the chapter mainly focuses on the estimation of sugar cane yields in Southeastern Brazil by using spatial tools which have been integrated in ILWIS 3.7.1, open source software. This study underpins that the NDVI data along with the other meteorological data is of immense use for the estimation of crop yields. This gives a business orientation on the utility of spatial tools, but also has a livelihood implication where small scale farmers or out growers are involved in sugarcane production. Interestingly, sugarcane is a major crop in all the case study countries in this chapter.

The last section of the chapter also gives more emphasis on yield estimates of five major crops in Uganda. The results of the study showed that the production between 2001 and 2010 has modestly increased with the variability in yields. Also, this analysis showed that the extensification of crops is dominated by intensification and it is implied that the increments in production are related to extensification.

In a nutshell, the chapter demonstrates how remotely sensed data available in the public domain freely or at very low cost can be harnessed to address critical challenges in developing countries pertaining to environment, agricultural productivity, drought, desertification and ultimately climate change adaptation. The chapter shows that relating the satellite derived vegetation indices with existing models and parameters can be useful proxies to understand the various phenomena of the crops. However, despite the availability of the technology, full benefits from available remotely sensed imagery resources for developing countries can only be realized when enabling policies are formulated and implemented and concerted capacity development is undertaken to establish a critical human resource base. This will enable the policy makers to go for the risk managing practices such as agricultural crop reinsurance schemes, drought defining criteria etc.

In light of the resource constraints in developing countries, cooperation and collaboration is important to develop a nucleus of future demand and contributing to new scientific insights related to projected changes in drought drawing information from satellite data, which will contribute to the improvement of warning systems and adaptation measures in developing and emerging countries.

REFERENCES

1. L Wolf, and M Williams, GEONETCast- Delivering environmental data to users worldwide (September 2007IEEE Systems Journal 2008;23401405
2. UNEPClimate change strategy, United Nations Environment Programme 2008
3. IPCCClimate Change 2007Synthesis Report- Contribution of Working Groups I, II and III to the Fourth Assessment Report of the Intergovernmental Panel on Climate Change, Geneva, Switzerland 2007.
4. G Nelson, M Rosegrant, J Koo, R Robertson, T Sulser, T Zhu, C Ringler, S Msangi, A Palazzo, M Batka, M Magalhaes, R Valmonte-santos, M Ewing, and D Lee, Climate Change: Impact on Agriculture and Cost of Adaptation," IFPRI, 200919
5. Barbosa H.A; Lakshmi Kumar2011Strengthening regional capacities for providing remote sensing decision support in drylands in the context of climate variability & change. In: Young, S. S.; Silvern, S.E. Environmental change. USA, 979-9-53307-109-0
6. C Vörösmarty, P Green, J Salisbury, and R Lammers, Global water resources: vulnerability from climate change and population growth. Science 2000289284
7. WMO2011Climate Knowledge for Action: A Global Framework for Climate Service- Empowering the Most Vulnerable. World Meteorological Organization, WMO-1065
8. P Sellers, Canopy reflectance, photosynthesis and transpiration. International Journal of Remote Sensing 19856813351372
9. Sarma A.A.L.N & Lakshmi Kumar, T.2006*Studies on crop growing period and NDVI in relation to water balance components*, Indian Journal of Radio & Space Physics, 35, 424-434.
10. C Bhuiyan, F. N Kogan, 2008*Monsoon dynamics and vegetative drought patterns in the Luni basin under rain-shadow zone,* Internation Journal of Remote Sensing, 3132233242
11. Lakshmi Kumar T.V., Humberto A. Barbosa, K. Koteswara Rao & Emily Prabha Jothi(2012*Studies on the frequency of extreme weather events over India*, Journal of Agriculture Science & Technology, 1413431356

12. J Cihlar, St. Laurent, L., & Dyer, J.A., (1991*The relation between normalized difference vegetation index and ecological variables,* Remote sensing of Environment, 35279298

13. M. L Davenport, S. E Nicholson, 1993*On the relation between rainfall and the Normalized Difference Vegetation Index for diverse vegetation types in East Africa,* International Journal of Remote Sensing, 14, 2369.

14. H. A Barbosa, M. d. S. & T. V Mesquita, Lakshmi Kumar (2011*What do vegetation indices tell us about the dynamics of the Amazon evergreen forests?,* Geophysical Research Abstracts, 13, EGU 2011, 12894.

15. F. N Kogan, 1997*Global Drought Watch From Space.* Bulletin of the American Meteorological Society, 78, 621636

16. L. S Unganai, and F. N Kogan, 1998*Southern Africa's Recent Droughts From Space.*Advance in Space Research, 21, 3507511

17. P. S Ramesh, R Sudipa, and F Kogan, 2003*Vegetation and temperature condition indices from NOAA AVHRR data for drought monitoring over India.* International Journal of Remote Sensing, 24, 43934402

18. Z Wan, P Wang, and X Li, 2004*Using MODIS land surface temperature & normalized vegetation index for monitoring drought in the sourthern Great Plains, USA*, Int. J. Remote Sens., 256172

19. J. K Knight, R. L Lunneta, J Ediriwickerma, S Khorram, 2006*Regional scale land cover characterization using MODIS- NDVI. 250 m multi-temporal imagery: A phenology based approach.* GIScience and Remote Sensing (Special issue on Multi- Temporal Imagery Analysis), 43(1), 1-23

20. Chris Funk & Michale E. Budde (2009*Phenologically- tuned MODIS NDVI based production anomaly estimates for Zimbabwe.* Remote Sensing Environment. 113115125

21. F Lupo, I Reginster, and E Lambin, Monitoring land-cover changes in West Africa with SPOT Vegetation: impact of natural disasters in 1998-1999. International Journal of Remote Sensing 2001221326332639

22. W Immerzeel, R Quiroz, and S De Jong, Understanding precipitation patterns and land use interaction in Tibet using harmonic analysis of SPOT VGT-S10 NDVI time series. International Journal of Remote Sensing 2005261122812296

23. J Xiao, and A Moody, Geographical distribution of global greening trends and their climatic correlates: 1982-1998. International Journal of Remote Sensing 2005261123712390
24. A Anyamba, and C Tucker, Analysis of Sahelian vegetation dynamics using NOAA-AVHRR NDVI data from 1981-2003. Journal of Arid Environments 2005633596614
25. U Helldén, and C Tottrup, Regional desertification: a global synthesis. Global and Planetary Change 200864169
26. Z Bai, D Dent, L Olsson, and M Schaepman, Global Assessment of Land Degradation and Improvement- 1. Identification by Remote Sensing, Report 2008/01, ISRIC- World Soil Information, Wageningen 2008
27. E Lepers, E Lambin, A Janetos, R De Fries, F Achard, N Ramankutty, and R Scholes, A synthesis of information on rapid land-cover change for the period 1981-2000. BioScience 2005552115124
28. G Baldi, M Nosetto, R Aragón, F Aversa, J Paruelo, and E Jobbágy, Long-term satellite NDVI data sets: evaluating their ability to detect ecosystem functional changes in South America. Sensors 200885397
29. F Achard, H Eva, H-J Stibig, P Mayaux, J Gallego, T Richards, and J-P Malingreau, Determination of deforestation rates of the world's humid tropical forests. Science 2002297999
30. UNCCDDesertification: a visual synthesis, United Nations Convention to Combat Desertification (UNCCD) Secretariat, Bonn 2011
31. C Tote, K Beringhs, E Swinnen, and G Govers, Monitoring environmental change in the Andes based on SPOT-VGT and NOAA-AVHRR time series analysis. In 6th International workshop on the Analysis of Multi-Temporal Remote Sensing Images, 2011
32. A Klisch, A Royer, C Lazar, B Baruth, and G Genovese, Extraction of phenological parameters from temporally smoothed vegetation indices. 2006
33. Z Bai, and D Dent, Global assessment of land degradation and improvement: pilot study in Kenya, Report 2006/01, ISRIC- World Soil Information, Wageningen 2006
34. A Peters, E Walter-shea, L Ji, A Vina, M Hayes, and M Svoboda, Drought monitoring with NDVI-based standardized vegetation

index. Photogrammetric Engineering and Remote Sensing 20026817175

35. N Guttman, Accepting the standardized precipitation index: a calculation algorithm. Journal of the American Water Resources Association 1999352311322
36. A Lotsch, M Friedl, B Anderson, and C Tucker, Coupled vegetation-precipitation variability observed from satellite and climate records. Geophysical research Letters 2003
37. Z Bai, and D Dent, Land Degradation and Improvement in Argentina- 1. Identification by remote sensing, ISRIC- World Soil Information, Wageningen 2007Thoronthwaite and Mather, 1955)
38. Lakshmi Kumar T.V., Uma, R, Aruna, K & Emily Prabha Jothi(2011On the relation of NDVI with rainfall in Western Ghats, India, RAISE Conference Proceedings, 4245Dec 2011.
39. Lakshmi Kumar T.V., Uma, R, Koteswara Rao K & Humberto Barbosa(2012Variability of NDVI in relation to southwest monsoon over Western Ghats, India, International SWAT Conference, July 2012.
40. Liu Xiaomang Dan Zhang., YuzhouLuo&Changming Liu(2012Spatial and temporal changes in aridity index in northwest China: 1960 to 2010, Theoretical and Applied Climatology, DOIs00704-012-0734-7, Aug 2012.
41. www.indiawaterportal.org
42. Thoronthwate C W & Mather J.R(1955The water balance model, Publ in Clim, Drexel Institute, 8,1, 104
43. C W Thoronthwaite, 1948An approach toward a rational classification of climate, Geographical Review, 38(1), 55-94.
44. J. R. F Gouvea, 2008Mudancas Climaticas e a expextiva de seus impactos da cultura de cana- de acucar na regiao Piracicaba-*SP,* Dissertacao (Mestrado em Fisica do Ambiente Agricola)- Escola Superior de Agricultura " Luiz de Querioz", Universidade de Sao Paulo, Piracciaba, 99.
45. B. F. T Rudorff, 1985Dados Landsat na estimativa da produtividade agrícola da cana-de- açúcar.114, Dissertação (Mestrado em Sensoriamento Remoto)- Instituto Nacional de Pesquisas Espaciais-INPE, São José dos Campos.
46. J Doorenbos, A. H Kassam, 1979Yield response to water. Roma: Food and Agriculture Organization of the United Nations, 193.

47. J. C. J Munoz, J. A Sobrino, L Guanter, J Moreno, A Plaza, P Matinez, 2005Fractional Vegetation Cover Estimation from Proba/CHRIS Data:Methods, Analysis of Angular Effects and Application to the Land Surface Emissivity Retrieval. 3rd ESA CHRIS/PROBA Workshop, ESRIN, ESA SP-593, Frascati, Italy.
48. J. M Norman, M. C Anderson, W. P Kustas, A. N French, J Meicikalski, R Torn, G. R Daik, T. J Achmugge, Remote Sensing of Evapotranspiration for Precision-Farming Applications. In: IEEE International Geoscience and Remote Sensing Symposium, 2003Tolouse. Proceedings: IGARSS 2003OrganizingCommittee, 21-25.
49. L. M. S Berka, B. F. T Rudorff, Y. E Shimabukuro, Soybean yield estimation by an agrometeorological model in a GIS. Scientia Agricola, 60n. 3; 4334402003
50. Toledo Filho M. R.1988Probabilidade de suprimento da demanda hídrica ideal da cultura de cana- de-açúcar (Saccharum spp.) através da precipitação pluvial na zona canavieira do estado de Alagoas. 1988.72f. Dissertação (Mestrado)- Progama de Pós-graduação em Agronomia, Escola Superior de Agricultura "Luiz de Queiroz", Universidade de São Paulo, Piracicaba.
51. Barbosa H. A; Rocha, D; Lakshmi Kumar, T.V., Bamutaze, Y.Quantification of sugarcane crop productivity using agrometeorological-spectral model: A study case in Northeastern Brazil. Revista AGROLLANIA, 1690-80669Enero-Diciembre, 62-75, 2012
52. http://www.earthobservations.org/geonetcast.shtml,
53. http://www.eumetsat.int/Home/Main/DataAccess/EUMETCast/SP_20100519114624675?l=en,
54. http://www.itc.nl/Pub/WRS/WRS-GEONETCast/GEONETCast-toolbox.html
55. http://faostat.fao.org/